TELEPHONE MEDICINE

For a catalogue of publications available from ACP-ASIM, contact:

Customer Service Center
American College of Physicians–American Society of Internal Medicine
190 N. Independence Mall West
Philadelphia, PA 19106-1572
215-351-2600
800-523-1546, ext. 2600

Visit our Web site at www.acponline.org

TELEPHONE MEDICINE

A GUIDE FOR THE PRACTICING PHYSICIAN

Anna B. Reisman

Assistant Professor
Yale University School of Medicine

David L. Stevens

Clinical Assistant Professor
New York University School of Medicine

AMERICAN COLLEGE OF PHYSICIANS

PHILADELPHIA, PENNSYLVANIA

Clinical Consultant: David R. Goldmann, MD
Manager, Book Publishing: David Myers
Production Supervisor: Allan S. Kleinberg
Production Editor: Karen C. Nolan
Acquisitions Editor: Mary K. Ruff
Developmental Editor: Vicki Hoenigke
Editorial Coordinator: Alicia Dillihay
Interior and Cover Design: Kate Nichols
Indexer: Nelle Garrecht

Printed in the United States of America
Composition by UB Communications
Printing/binding by Versa Press

American College of Physicians (ACP) became an imprint of the American College of Physicians–American Society of Internal Medicine in July 1998.

Library of Congress Cataloging-in-Publication Data

American College of Physicians–American Society of Internal Medicine.
    Telephone Medicine: A Guide for the Practicing Physician / [edited by] Anna B. Reisman, David L. Stevens–1st ed.
        p. ; cm.
    Includes bibliographical references and index.
    ISBN 0-943126-87-8
    [DNLM: WX 173 E383 2000]
R864.A42 2000
651.5'04261'0285–dc21

The authors and publisher have exerted every effort to ensure that drug selection and dosage set forth in this manual are in accord with current recommendations and practice at the time of publication. In view of ongoing research, occasional changes in government regulations, and the constant flow of information relating to drug therapy and drug reactions, the reader is urged to check the package insert for each drug for any change in indications and dosage and for added warnings and precautions. This care is particularly important when the recommended agent is a new or infrequently used drug.

02 03 04 05 06 / 9 8 7 6 5 4 3 2 1

With love and gratitude to
Pam, Josie, and William;
and Cary and Jacob;
for their unflappable support

CONTRIBUTORS

Ayse A. Atasoylu, MD, MPH
Instructor
Harvard Medical School;
Attending Physician
Cambridge Hospital
Cambridge, Massachusetts

Sary O. Beidas, MD, FACP
Assistant Professor
Howard University School of Medicine;
Program Director
Internal Medicine Residency
Prince George's Hospital Center
Cheverly, Maryland

Ira Daniel Breite, MD
Clinical Instructor
Department of Medicine
Division of Gastroenterology
New York University School of
    Medicine;
Westside Medical Associates LLP
New York, New York

Michael S. Cohen, MD
Clinical Instructor
Department of Dermatology
New York University School of
    Medicine
New York, New York

Peter Curtis, MD
Professor
Department of Family Medicine
University of North Carolina
    School of Medicine
Chapel Hill, North Carolina

Cary P. Gross, MD
Assistant Professor
Department of Internal Medicine
Yale University School of
    Medicine
New Haven, Connecticut

Sally G. Haskell, MD
Assistant Professor
Section of General Internal Medicine
Yale University School of Medicine;
Director, Women's Clinic
Connecticut VA Healthcare Center
West Haven, Connecticut

Mack Lipkin, Jr., MD
Professor of Clinical Medicine
New York University School of
    Medicine;
Director, Division of Primary Care
Bellevue Hospital
New York, New York

Timothy S. Loo, MD
Clinical Instructor, Department of
    Internal Medicine
Johns Hopkins University School of
    Medicine;
Lead Physician, East Baltimore
    Medical Center
Baltimore, Maryland

Jeanne McCauley, MD, MPH
Assistant Professor
Department of Medicine
Johns Hopkins School of Medicine;
Medical Director, Clinical Research
    and Outcomes
Johns Hopkins Community
    Physicians
Baltimore, Maryland

Kei Mukohara, MD
Department of General Medicine
Nagoya University Hospital
Nagoya, Japan

Daniel H. Pomerantz, MD
Assistant Professor
Department of Medicine
New York Medical College
Sound Shore Medical Center of
    Westchester
New Rochelle, New York

Anna B. Reisman, MD
Assistant Professor
Department of Internal Medicine
Section of General Internal
    Medicine
Yale University School of Medicine;
Attending Physician
VA Connecticut Healthcare System
West Haven, Connecticut

Lisa Reisman, Esq.
Columbia University School of the
    Arts
New York, New York

William H. Salazar, MD
Associate Professor of Medicine and
  Psychiatry
Medical College of Georgia;
Associate Program Director
Internal Medicine Residency
  Program
Augusta, Georgia

Mark Schwartz, MD, FACP
Assistant Professor of Clinical
  Medicine
Department of Internal Medicine
New York University School of
  Medicine;
Director, Fellowship Program in
  General Internal Medicine;
Society Master
Walter Reed Society for Health
  Policy and Public Health;
Attending Physician
VA New York Harbor Healthcare
  System
New York, New York

David L. Stevens, MD
Clinical Assistant Professor
Department of Internal Medicine
New York University School of
  Medicine;
Attending Physician
Gouverneur Hospital
New York, New York

Sondra Zabar, MD
Clinical Assistant Professor
Department of Medicine
New York University School of
  Medicine;
Co-Director
Primary Care Residency Program;
Attending Physician
Gouverneur Hospital
New York, New York

FOREWORD

When a patient notices a new or worsening symptom, the symptom predictably awakens a cascade of concerns. The patient's first reaction usually involves tending to the symptoms through self-care. On any day, in communities throughout the world, approximately 30% of people are engaged in self-care, which can be defined as care that does not include contact with a health professional.

When the patient feels self-care is ineffectual, however, nowadays the first contact with a health professional is often a telephone call. The caller expects that the problem will be understood, addressed, and solved. This "expectant hope," as Jerome Frank called it many years ago, is every patient's reason for seeking professional help from a physician known to possess a sizable fund of knowledge. The strength of *Telephone Medicine: A Guide for the Practicing Physician* lies in its attention to what is unique as well as what is generic when the context of care is a telephone call.

I have often wished, and have probably told this to both of the editors, that I could visit each of my patients at home at least once. This wish reminds me of a unique feature of telephone medicine: my patients are usually in very familiar surroundings during our telephone encounters. And this is true for up to 25% of my encounters with them. Does being in familiar surroundings preserve some measure of comfort that is lost in my office or in an emergency department? During a telephone encounter, might a patient develop more confidence about the next steps we have agreed to because the setting for taking these steps is present and visible during our encounter? I, on my part, relinquish some measure of control over the situation when I cannot examine a patient as part of addressing symptoms. On balance, does this mix of situations that I and the patient find ourselves in usually work in the patient's favor? I would say "yes," if the physician has the training to care for patients by telephone. Here, the editors of *Telephone Medicine* have assembled the type of information needed by the physician who would acquire and utilize that expertise.

The pervasive message in this book is that telephone medicine is strictly communication medicine. The time available for communication is usually short. The physician's skills need to be both efficient and patient centered, considering the independence required of a patient for handling a problem at home. The patient's story, including the emotions attached to it, must allow the internist to locate the illness in a spectrum ranging from low to high risk, and the physician's response must suffice for the patient to feel comforted, to comprehend the nature of the illness and the physician's instructions, and to take appropriate actions.

The first section of *Telephone Medicine* grounds the reader in general considerations. Next, the clinical section covers the most common and important patient problems. These 13 chapters are an excellent resource to consult when one needs practical, evidence-based information in a telephone encounter. A final chapter in the clinical section discusses difficult patients and how to deal with them.

I have long advocated demystifying telephone medicine and making it part of training. The final section, which discusses how to incorporate telephone media into the workplace, has as its final chapter a tested curriculum that nicely closes the loop for those of us who have said of many aspects of doctoring that we need an effective way to help residents learn.

A test of the usefulness of *Telephone Medicine* will be how it affects the practicing physicians and students who read it. I predict that there will be favorable changes in their methods of telephone communication and diagnosis, and that their patients will be much better off when they telephone their physicians.

L. Randol Barker, MD, ScM
Professor of Medicine, Johns Hopkins University School of Medicine
Co-Director, Division of General Internal Medicine, Johns Hopkins Bayview
*   Medical Center, Baltimore, Maryland*

PREFACE

Why a book on telephone medicine? The telephone is now a significant component of medical care: 25% of internists' encounters with patients take place over the telephone. The uses of the telephone are many: to clarify issues raised at the office visit, to follow up on patients recently discharged from the hospital, to help patients with decisions about their health care at home, to prevent unnecessary emergency visits, to give test results, and to provide reassurance for both patient and physician. New medical problems may be diagnosed earlier, and some chronic diseases may be managed. Telephone medicine can also improve the rapport between doctor and patient, increase access to care, enhance patient satisfaction, and, in some cases, lower patient and physician costs of care.

However, despite these benefits, telephone medicine is not without its difficulties. Many physicians are uncomfortable talking with patients on the telephone, and most physicians have not been trained in telephone medicine. Additionally, a physician's impression of a patient's status is largely based on visual information, and this is obviously lacking on the telephone. The communication skills required for effective encounters on the telephone are different than those used in the office setting. Listening skills become even more important when providing care over the telephone because of the absence of visual cues.

Telephone Medicine: A Guide for the Practicing Physician addresses these challenges. A practical, evidence-based guide for the internist, it provides information on telephone communication skills, documentation and medicolegal aspects of telephone medicine, the telephone in managed care, managing the difficult patient over the telephone, office management issues, and teaching telephone medicine, in addition to offering a strong focus on evidence-based approaches to the telephone management of common symptoms.

Telephone Medicine will be useful to physicians working in private practice, managed care, and academia. We hope you find it helpful in your practice of medicine.

Anna B. Reisman
David L. Stevens

ACKNOWLEDGEMENTS

We wish to express our thanks to the following people for their comments on individual chapters: Andrew Lowe, Rick Haeseler, William Slater, Daniel Federman, and Isaac Silverman. We also thank many of the primary care and internal medicine residents at New York University, Johns Hopkins University, and Yale for feedback on our teaching of telephone medicine.

CONTENTS

III. INCORPORATING TELEPHONE MEDICINE INTO THE WORKPLACE

PART I

....

PRACTICING TELEPHONE MEDICINE: GENERAL APPROACH TO THE TELEPHONE INTERVIEW

1

OVERVIEW OF TELEPHONE MEDICINE

Anna B. Reisman, MD

*Internal medicine is an endless series of telephone calls interrupted
by occasional live patients who happen to wander into one's office.*
MICHAEL J. HALBERSTAM (1)

The telephone is a significant tool in the practice of medicine. It is
estimated that approximately 25% of internists' interactions with
patients are over the phone, a number that may be increasing
with managed care's emphasis on promoting telephone access (2-4).
Telephone contact has the potential to improve clinical outcomes and patient satisfaction as an adjunct to office practice and to enhance the
doctor-patient relationship. Telephone access to physicians can also
reduce emergency department use. One recent study found that 33% of
patients would have gone to the emergency department if they had been
unable to reach a physician; only 8% went to the emergency department
after telephone contact with a physician (5) (Figure 1-1).

Effective telephone medicine requires a refinement of certain com-
municative skills and knowledge. The lack of a physical exam is one feature of telephone medicine that distinguishes it from office medicine and
renders it especially challenging. The physician, without seeing the patient and without access to the patient's record, must make potentially
critical decisions about whether the patient needs to be seen as an emergency, whether to prescribe a medication, or whether to opt for watchful
waiting.

Physicians are less effective when they practice over the phone than
when they have face-to-face interviews with their patients. Not surprisingly,

one study found that most practicing internists did not have confidence in their telephone medicine abilities. The quality of telephone medicine practice by physicians has been called into question by several articles in the pediatric literature (6,7). A 1978 study reported that nurse practitioners performed telephone medicine more effectively and ascribed this to the paucity of telephone skills training for pediatricians (6). Although a number of studies have shown that training in telephone medicine skills can improve physician attitudes (8), few internal medicine residency programs teach telephone medicine. Most internists learn how to practice telephone medicine by trial and error.

HISTORY OF TELEPHONE MEDICINE

Since its inception in the late 1870s, the telephone has played a key role in the practice of medicine. Indeed, the first telephone exchange on record connected a Connecticut drugstore with 21 local physicians (Starr P. The Social Transformation of American Medicine. New York: Basic Books; 1982). Physicians quickly realized the practice-boosting potential of the telephone. In 1878, a Californian physician placed an advertisement in a newspaper stating his availability by phone, day or night (Reiser S. Medicine and the Reign of Technology. New York: Cambridge University Press; 1978). Physicians were the first professionals to have telephones and to be "on call" around the clock (Aronson S. The Lancet on the telephone: 1876-1975. Medical History. 1977;21:69-87).

This was, however, the era known as the "age of physical diagnosis," and many physicians did not believe in making a diagnosis over the telephone. Some realized that the telephone could transmit heart and lung sounds over long distances, but these experiments were largely unsuccessful (Reiser). Nonetheless, the telephone's use as a diagnostic tool grew. In 1879, a Cincinnati physician listened to a child's cough on the telephone and determined that the cause did not warrant a house call: the child was fine when he saw him the next morning (Reiser).

This increased access to physicians through the telephone gained popularity. By the late 1800s, the telephone had become a nuisance and even a compulsion to physicians who were responding to calls even while in the midst of examining a patient (Aronson). One writer described the physician as "a slave to the peremptory call of the instrument [who] must constantly be ready to answer it, as in many cases no one can answer it for him." (Reiser; Hildreth J. The general practitioner and the specialist. Boston Med Surg J. 1906;155:79). To protect physicians from the intrusion of nonserious cases, The Lancet suggested an etiquette for telephone medicine that calls should be limited to true urgencies; however, it was often difficult to define exactly what qualified as an "urgency" (Aronson). Eventually, the concept of having an assistant to screen and prioritize calls came into being (Aronson).

Operator-assisted calls and party lines left many callers worried about providing personal information on the phone. Issues of confidentiality and privacy were paramount in telephone medicine until the advent of private telephone lines in the middle of the 20th century (Starr).

The number of homes with telephones has continued to rise: In 1994, 97.4 million households in the United States had telephones (Rickert R. The Telephone. In: Microsoft Encarta Encyclopedia, 1998 ed.). An estimated 34.3% of Americans own cellular phones (Murphy D. Two continents, disconnected. The New York Times. December 14, 2000; B1, B6). All physician practices have a telephone system of one sort or another, many with nurse triage as the front line, and most with a system for after-hours calls. More than a century after the advent of the telephone, many of the challenging issues of early telephone medicine (time management, privacy, the difficulty of making diagnoses over the telephone, and triage) remain at the forefront today.

PATIENT AND PHYSICIAN PERSPECTIVES ON TELEPHONE MEDICINE

Patient Perspective

Most patients are generally satisfied with the telephone care they receive (5,9). One study of telephone medicine outcomes assessed patient satisfaction with phone interaction 1 week later. Two-thirds of patients ranked their

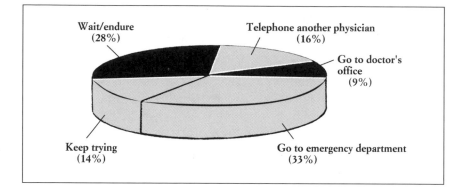

Figure 1–1 Responses of patients to the question, "What would you have done if you were not able to reach a physician by telephone?" (Modified from Delichatsios H, Callahan M, Charlson M. Outcomes of telephone medicine. J Gen Intern Med. 1998;13:583.)

level of satisfaction as excellent or very good, and 21% as good. Those most satisfied tended to have complete relief of their symptoms (5).

Common reasons for dissatisfaction with telephone contacts include physician refusal to prescribe medication, lack of access to one's regular physician, and a recommendation that the patient do what he or she was already doing (9).

For many patients, reassurance may be more important than symptom relief (10). Often, it is not the actual physical symptoms that prompt the patient to call, but rather the worry that the symptoms might represent something dangerous. In one study where patients were surveyed about what they considered the most important element of an after-hours telephone encounter, 49% reported that reassurance was more important than relief of symptoms (10).

Physician Perspective

Physicians are generally dissatisfied with their performance in telephone encounters (11,12), reporting discomfort in prescribing medications over the phone (especially to patients seeking narcotics and patients needing antidepressants). Almost a third of physician contacts over the telephone produce indifference, frustration, and anger (10). Nonetheless, physicians view most telephone calls as reasonable or appropriate (10,13,14), although patients view a higher number of their calls as "true emergencies" than do physicians (9,11,15). In almost one-third of the calls, the physician and patient may see the main reason for the call differently (10); in one study, physicians tended to perceive the reason for most calls as physical problems, whereas callers expressed their main reason for calling as anxiety or concern (15).

Despite the significant proportion of time spent by physicians on the telephone, many physicians report feeling a lack of confidence in their abilities on the phone. Most physicians have received no training in telephone medicine. Only 6% of residency programs formally taught telephone medicine in 1995 (16). In this respect, telephone medicine is no different from other areas in which physicians practice what they have not been traditionally trained to do, such as managing chronic pain or diagnosing and treating depression.

The paucity of publications on telephone medicine for the practicing internist further heightens the challenge of telephone medicine. There are no expert panel guidelines and few controlled trials. Although guides for telephone medicine exist in the pediatric and nursing literature (7,17-19), there are no current books on the subject specifically designed for the practicing internist. The literature on telephone medicine in internal medicine, compared with the family practice and pediatric literature, is sparse.

The current state of telephone medicine is not, however, as bad as it seems. The medical history is paramount in the practice of effective telephone medicine, where information from the physical examination and laboratory testing is usually lacking; indeed, many diagnoses in internal medicine are made based primarily on the medical history. Until recently, there has been little scientific examination of which specific elements in the medical history help with accurate diagnosis. Much of what has been learned about the practice of medicine in recent years is applicable to the care of the patient over the phone. Specifically, evidence-based analysis of the utility of the medical history and advances in our understanding of the medical interview are both directly relevant to telephone medicine. The primary goals of *Telephone Medicine: A Practical Guide for the Physician* are therefore 1) to produce evidence-based guidelines for the management of medical problems commonly encountered on the phone and 2) to summarize our understanding of the medical interview as it applies to telephone encounters in general and as it applies to medical and psychosocial situations specifically.

OVERVIEW OF CHAPTERS

The Telephone Interview

What makes the telephone interview particularly challenging? A well-known model of the medical interview defines its three functions as information gathering, developing rapport, and educating the patient (20). Each function poses different challenges when applied to the telephone interview.

Information Gathering

The lack of visual cues from the patient over the telephone makes information gathering more difficult. The physician does not see the bouncing knee and gnawed fingernails of an anxious patient, the straining sternocleidomastoids of a patient with an acute asthma attack, the avoidance of eye contact in a patient with depression, or the grimace of a patient in pain.

Developing Rapport

The development of rapport with a patient is especially challenging over the telephone. Time is usually limited, and it is less likely that there will be a pre-existing doctor-patient relationship. Simple nonverbal cues and actions by the physician that often help create rapport are often not possible over the telephone. For example, the patient cannot see the physician's open posture that might convey empathy and a willingness to listen, nor can he appreciate the handshake and eye contact that might indicate the physician's interest and attention.

Educating the Patient

Similarly, the paucity of nonverbal cues also challenges effective patient education over the telephone. The physician cannot see, for example, the blank stare of a patient that might signal a lack of comprehension, and the patient may not feel adequately reassured without seeing the physician's confident and caring demeanor.

Time limits and outside distractions also make patient education over the telephone difficult. The physician who receives a call at the office, for example, may feel the pressure of patients waiting; the physician at home may be trying to get back to a family dinner. A patient may feel like an imposition when speaking with the physician, and physicians may feel rushed to terminate the call to return to previous activities. The brevity of most telephone encounters leaves less time for a discussion of the plan, and busy or distracted physicians may cut short patient education once he or she is satisfied that there is no emergency. The lack of written instructions for the patient may compound this problem.

Chapter 2 presents more detailed information on these aspects of the telephone interview. For each function of the medical interview, an annotated list of essential elements and necessary skills is provided. Strategies for eliciting information from patients are provided, with a focus on finding hidden concerns. Discussion of case examples provides a general approach to improving communication over the telephone, as well as ways to deal with specific communication barriers.

Medicolegal Issues of Telephone Medicine

An understanding of medicolegal issues relevant to practicing medicine over the telephone is paramount for all physicians. Physicians should be familiar with what types of interactions constitute a doctor-patient relationship, as well as the definition of medical malpractice. Chapter 3 also describes when it is appropriate, and when it is not, to give medical advice over the phone. Issues such as the proper way to deal with disconnected calls, unreachable patients, and medication questions are discussed, as well as broader issues such as confidentiality and competency.

Clinical Chapters

The clinical chapters in this book (Chapters 4 to 16) focus on common symptoms that are reported over the phone, with a focus on acute symptoms. An approach to the telephone management of each symptom is presented in a structured series of questions and answers. The evidence on the efficacy of the medical history regarding each symptom was gathered, analyzed, and applied, when possible, to the telephone. Because few studies have been done on the telephone management of most symptoms, the evidence has been culled from epidemiological data and studies that focus on using aspects of the history as a means of diagnosis.

The focus in these chapters is on management rather than diagnosis, for two reasons. First, definitive diagnosis is often impossible without physical exam and lab tests. Second, the primary task with many callers is triage: whether this caller needs to be seen in person and when. If in-person evaluation is not needed, management usually focuses on reassurance and symptom relief.

Difficult Patients

Many physicians feel most challenged on the telephone when faced with a difficult patient. Although some difficult patients fall into the well-known categories of the personality disorders, such as the dependent patient or the histrionic patient, others present difficulty more from their specific telephone-related behaviors. One patient may be angry; another may use the telephone exclusively and never come in for an office visit; and another may inappropriately call the physician at home.

In a case-based discussion format, Chapter 17 provides guidelines for managing specific difficult patient behaviors and suggested strategies for dealing with common personality styles that may manifest themselves over the phone.

Electronic Advances

E-mail and the Internet are changing the face of medicine. Like the telephone, they provide an alternative way for doctors to communicate with patients. Some physicians prefer e-mail to the telephone because they can respond to e-mail in their own time, and because the text of an e-mail exchange can be printed out and saved for the medical record. The Internet provides a way for physicians to offer information to patients, and it is also a means by which patients can educate themselves on medical issues.

There are, however, many potential problems with these technologies. Patients who are not clear about the appropriate use of e-mail may send a message about an urgent or emergency problem that is not immediately read by the physician. If encryption is not effective, there can be a loss of privacy. Additionally, not all information on the Internet is accurate. Chapter 18 focuses on these and other issues related to emerging technologies such as telemedicine and interactive telephone systems.

Telephone Medicine in the Office

More recent developments in the office practice of telephone medicine, from managed care to direct marketing campaigns, have increased the number of calls received from patients. Chapter 19 provides information that will be useful in understanding and addressing some of these challenges. These include telephone triage systems, after-hours services, fees for telephone consultations, telephone policies, physician availability, emergency department authorization, and prescription refills.

Teaching Telephone Medicine

As mentioned above, only 6% of residency programs teach telephone medicine. This lack of training probably plays a significant role in physicians' lack of comfort with practicing medicine over the telephone. In Chapter 20, a brief curriculum for teaching skills in telephone medicine to students, residents, and practicing physicians is presented, along with role play scenarios and a script for a trigger tape. The curriculum focuses on effective communication skills, completeness of the medical history, identification of hidden patient concerns, and documentation.

BENEFITS AND USES OF TELEPHONE MEDICINE

The benefits of telephone medicine go far beyond the management of an acute problem. There are many reasons why a physician might speak to a patient on the phone. Some of these have been studied and found to be effective, and the individual physician can find many novel ways to incorporate

---

Table 1-1   Uses and Benefits of Telephone Medicine

Uses
- Augment issues raised at the office visit
- Give patients help with health care decisions at home
- Modify treatment for chronic disease (e.g., hypertension, diabetes mellitus) based on home monitoring
- Refill medications
- Give test results
- Diagnose problems early
- Prevent unnecessary emergency department visits
- Follow up patients discharged from hospital
- Allow patients to speak with physician from more comfortable environment
- Direct family members in CPR
- Provide information on suicide hotlines and counseling

Benefits
- Improved doctor-patient rapport
- Improved access to care
- Increased patient satisfaction
- Cost-saving for patient
- Potential for reimbursement
- Improved efficiency in managed care setting
- Prevention of unnecessary emergency department visits
- Improved outcomes
- Avoidance of missed days of work
- Reduced demand on office staff

---

telephone communication into his or her practice. Table 1-1 lists some possible uses and benefits of the telephone, several of which are described in more detail below.

Access

Effective telephone access systems allow patients to contact their physicians for intercurrent problems, advice, and treatment. In rural areas, the telephone can provide vital and timesaving information. Patients with financial difficulties may seek assistance on the phone to help them determine if more costly office or emergency department visits are necessary. For working adults, a brief phone call to the physician requesting a recommendation for treatment of a minor illness may help the patient avoid missed days of work. Patients with small children may be saved the trouble of finding last-minute childcare. Patients can also use the telephone system to learn test results and request medication refills rather than waiting for their next appointment. For some anxious patients, the telephone provides a less stressful way to speak with the physician (21); such patients might otherwise avoid their regular follow-up visits.

Decreased Utilization of Medical Services

The improved access to the physician via telephone can decrease in-person utilization of medical services, leaving more time and resources for patients who require in person evaluation. In one study, patients receiving scheduled telephone calls in place of some scheduled follow-up visits had a reduced rate of hospitalization, medication use, and costs (22). Another study found an equal level of satisfaction in patients followed up by telephone and seen in person (23). Several studies have shown that fewer emergency department visits are made when patients have telephone access to a physician (5,15), although a more recent study that substituted telephone appointments for clinical visits found no reduction in the use of medical services (23a). In the era of managed care, many patients are being discharged from the hospital earlier than they would have been previously. This results in a higher proportion of sicker patients in the community; telephone follow-up of such patients can provide a convenient and cost-saving way to ensure clinical improvement.

Proactive Use of the Telephone

Telephone calls in internal medicine can be grouped as doctor-initiated calls, patient-initiated (or family/companion-initiated) calls, and calls initiated by other medical personnel. This book will focus on patient-initiated (or family/companion-initiated) calls, or reactive calls. These calls are made for new symptoms, administrative issues, and follow-up of chronic problems. Doctor-initiated calls, or proactive calls, are made by physicians for follow-up of acute issues relating to recent office or telephone encounters, follow-up of chronic issues such as blood pressure or blood glucose measurements, or to assess the efficacy of a new medication (Table 1-2).

Table 1-2   Types of Telephone Calls

Doctor-Initiated Calls (Proactive)

Follow-up of acute issue
  Examples:
  • Assess response to treatment such as antibiotics
  • Assess patients after hospital discharge
  • Lab results such as throat culture

Follow-up of chronic issue
  Examples:
  • Lab results such as lipid panel
  • Assess follow-through with plan such as smoking cessation
  • Check in with patient who has missed an appointment

Patient-Initiated Calls (Reactive)

New symptoms/urgent calls
  Examples:
  • Discomfort from new symptoms such as diarrhea
  • Anxiety from new symptoms such as new headache

Medication questions
  Examples:
  • Medication side effects
  • Concern that an over-the-counter medicine may interact with prescription
    medication

Administrative issues
  Examples:
  • Make/change appointment
  • Authorization to see another health care provider

Physicians or other health professionals can use the telephone to initiate calls to patients to improve outcomes. A number of randomized controlled trials have demonstrated the usefulness of proactive calls in patient care. Studies looking at telephone follow-up of cardiac care after surgery or myocardial infarction showed improvement in behaviors and outcomes such as exercise capacity, smoking cessation, low-density lipoprotein cholesterol lowering, and more rapid recovery to normal activity (24). Patients with osteoarthritis who were called regularly with a focus on promoting self-care had less pain and increased physical activity (25). Results from some studies of telephone follow-up, including issues of smoking cessation, adherence to continuous positive airway pressure for sleep apnea, and anxiety scores for patients undergoing radiation therapy, did not show a benefit (24).

EPIDEMIOLOGY OF TELEPHONE MEDICINE

Women call their physicians more often than men do, making approximately two-thirds of calls in both internal medicine and family practice settings (3,5,9,10,14,15,26-28). This number usually mirrors the practice population. Most calls from patients in internal medicine settings are made by older adults (29), although the emphasis by managed care on telephone access may be shifting the age of callers downward. Patients in managed care plans call their physicians more often than patients with other types of health care plans. Middle-class patients use the telephone more and make fewer appointments, whereas patients in lower socioeconomic groups often use the emergency department more than the telephone (30). Most clinical calls are initiated by the patient rather than by a family member (9,28).

Most patient-initiated calls are made during office hours (28,29), and most after-hours calls are made before midnight (9,13,31,32). The percentage of calls that are "serious" tends to increase as the day goes on, with the highest proportion during the late night (Figure 1-2) (32). The average duration of a phone call between physician and patient is less than 2 minutes

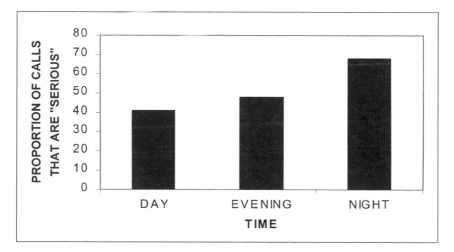

Figure 1-2 Relationship between serious calls and the time of day. (Modified from Starr P. The Social Transformation of American Medicine. New York: Basic Books; 1978.)

(14,28,29), although calls are longer when the problem is emotional and when the patient is not known to the physician (3).

Most calls during the day are for administrative issues, such as prescription refills, scheduling appointments, requests for referrals, and insurance questions, rather than symptoms (10,27). After-hours calls are usually made in reference to patient symptoms (28,32). Most symptom-related calls are for follow-up of known chronic or subacute problems (3,26,28).

The most common actions by physicians over the phone are reassurance and advice (3,28). In internal medicine, a recommendation to go the emergency department is made between 1% and 7% of the time (3,5,28,29), whereas reassurance alone is provided in 1/3 to 2/3 of phone calls (5,28). Other common dispositions include scheduling an earlier visit, giving a prescription, changing a dosage, counseling, and giving test results.

THE FUTURE OF TELEPHONE MEDICINE

What does the future hold for telephone medicine? The number of telephone calls between physicians and patients will likely continue to grow, prompting new developments with the goal of maximizing the efficacy and efficiency of calls while decreasing costs. Disease-management strategies, designed to lead to a greater degree of patient self-monitoring, may prompt more calls regarding the management of chronic illness. Cost-saving strategies such as the replacement of scheduled visits with scheduled phone calls may become more commonplace.

Telephone medicine will likely continue to merge with advances in technology such as telemedicine and the Internet. Although it is too early to tell whether telephone medicine will be gradually supplanted by these technologies or enhanced by them, the latter seems more likely. As new technologies expand physicians' abilities to prevent poor outcomes in conditions such as acute myocardial infarction and stroke with early interventions, the importance of appropriate telephone management will become even greater. The accuracy of telephone evaluation and intervention will depend on further research in the field and the continued development of evidence-based recommendations.

REFERENCES

1. **Halberstam M.** Medicine by telephone: is it brave, foolhardy, or just inescapable? Mo Med. 1977;74:11-5.
2. **Hallam L.** You've got a lot to answer for, Mr. Bell: a review of the use of the telephone in primary care. Fam Pract. 1989;6:47-57.
3. **Radecki S, Neville R, Girard R.** Telephone patient management by primary care physicians. Medical Care. 1989;27:817-22.
4. Medical Practice in the United States. Princeton, NJ: Robert Wood Johnson Foundation; 1981.
5. **Delichatsios H, Callahan M, Charlson M.** Outcomes of telephone medicine. J Gen Intern Med. 1998;13:579-85.
6. **Perrin E, Goodman H.** Telephone management of acute pediatric illnesses. N Engl J Med. 1978;298:130-5.
7. **Katz H.** Telephone Medicine Triage and Training: A Handbook for Primary Health Care Professionals Philadelphia: FA Davis; 1990.
8. **Fleming M, Skochelak S, Curtis P, Evens S.** Evaluating the effectiveness of a telephone medicine curriculum. Medical Care. 1988;26:211-6.
9. **Greenhouse D, Probst J.** After-hours telephone calls in a family practice residency: volume, seriousness, and patient satisfaction. Fam Med. 1995;27:525-30.
10. **Curtis P, Talbot A.** The after-hours call in family practice. J Fam Pract. 1979;9:901-9.
11. **Hannis M, Elnicki M, Morris D, Flannery M.** Can you hold please? How internal medicine residents deal with patient telephone calls. Am J Med Sci. 1994;308: 349-52.
12. **Hannis M, Hazard R, Rothschild M, et al.** Physician attitudes regarding telephone medicine J Gen Intern Med. 1996;11.678-83.
13. **Bergman J, Rosenblatt R.** After hours calls: a five-year longitudinal study in a family practice group. J Fam Pract. 1982;15:101-6.
14. **Daugird A, Spencer D.** Characteristics of patients who highly utilize telephone medical care in a private practice. J Fam Pract. 1989;29:59-64.
15. **Evens S, Curtis P, Talbot A, et al.** Characteristics and perceptions of after-hours callers. Fam Pract. 1985;2:10-6.
16. **Flannery M, Moses G, Cykert S, et al.** Telephone management training in internal medicine residencies: a national survey of program directors. Acad Med. 1995;70:1138-41.
17. **Wheeler S.** Telephone Triage: Theory, Practice and Protocol Development. Albany: Delmar Publishers; 1993.
18. **Brown J.** Pediatric Telephone Medicine: Principles, Triage, and Advice Philadelphia: JB Lippincott, Williams & Wilkins; 1994.
19. **Schmitt B.** Pediatric Telephone Advice, 2nd ed. Philadelphia: Lippincott-Raven; 1999.
20. **Lazare A, Putnam S, Lipkin M Jr.** Three Functions of the Medical Interview. In: Lipkin Jr MP, SM. Lazare, A, Eds. The Medical Interview: Clinical Care, Education, and Research. New York: Springer-Verlag; 1995.
21. **Mermelstein H, Holland J.** Psychotherapy by telephone: a therapeutic tool for cancer patients. Psychosomatics. 1991;32:407-12.

22. **Wasson J, Gaudette C, Whaley F, et al.** Telephone care as a substitute for routine clinic follow-up. JAMA. 1992;267:1788-93.
23. **Pal B.** Following up outpatients by telephone: pilot study. BMJ. 1998;316:1647.
23a. **Welch HG, Johnson DJ, Edson R.** Telephone care as an adjunct to routine medical follow-up. A negative randomized trial. Effect Clin Prac. 2000;3:123-30.
24. **Balas E, Jaffrey F, Kuperman G, et al.** Electronic communication with patients: evaluation of distance medicine technology. JAMA. 1997;278:152-9.
25. **Rene J, Weinberger M, Mazzuca S, et al.** Reduction of joint pain in patients with knee osteoarthritis who have received monthly telephone calls from lay personnel and whose medical treatment regimens have remained stable. Arthritis Rheum. 1992;35:11-5.
26. **Spencer D, Daugird A.** The nature and content of physician telephone calls in a private practice. J Fam Pract. 1988;27:201-5.
27. **Fischer P, Smith S.** The nature and management of telephone utilization in a family practice setting. J Fam Pract. 1979;8:321-7.
28. **Johnson B, Johnson C.** Telephone medicine: a general internal medicine experience. J Gen Intern Med. 1990;5:234-9.
29. **Morrison R, Arheart K, Rimner W.** Telephone medicine in a southern university private practice. Am J Med Sci. 1993;306:157-9.
30. **Weiss J, Greenlick M.** Determinants of medical care utilization: the effect of social class and distance on contacts with the medical care system. Med Care. 1970;8:456-62.
31. **Perkins A, Gagnon R, DeGruy F.** A comparison of after-hours telephone calls concerning ambulatory and nursing home patients. J Fam Pract. 1993;37:247-50.
32. **Peters R.** After-hours telephone calls to general and subspecialty internists: an observational study. J Gen Intern Med. 1994;9:554-7.
33. **Starr P.** The Social Transformation of American Medicine. New York: Basic Books; 1982.
34. **Reiser S.** Medicine and the Reign of Technology. New York: Cambridge University Press; 1978.
35. **Aronson S.** The Lancet on the telephone: 1876-1975. Medical History. 1977;21:69-87.
36. **Hildreth J.** The general practitioner and the specialist. Boston Med Surg J. 1906;155:79.
37. **Rickert R.** The telephone. In: Microsoft Encarta Encyclopedia; 1998 ed.
38. **Murphy D.** Two continents, disconnected. The New York Times. December 14, 2000; B1, B6.

2

. . . .

THE TELEPHONE INTERVIEW

. . . .

Peter Curtis, MD

*Over the phone it's a very lonely life indeed. "Do you think I need
to come in?" says the patient. You're busy, the patient has had this
particular complaint dozens of times before, the patient needs the
money more than you do. And yet this could be the time—the time
for the crocky chest pains to be an acute infarction, the time for the
'spastic colon' to be an acute obstruction, with gangrenous bowel
setting in, the time for "bleeding from my hemorrhoids" to be a
cancer on top of the hemorrhoids. There is no time to call in a committee. The decision must be made, and it is a consensus of one.*
MICHAEL J. HALBERSTAM (1)

A s outlined in the overview chapter (Chapter 1), the telephone interview is a major part of a physician's workday and a vital component of medical care. Nevertheless, several studies have
demonstrated that physicians more often show inadequacies and anxieties
in telephone interviews than they do in face-to-face encounters (2).

Physicians should be familiar with the essential components of a telephone interview and how the telephone interview differs from the in-
person encounter. An obvious but critical difference is that the telephone
evaluation relies solely on the spoken word for evaluation and for imple-
mentation of the management plan.

CHALLENGES IN THE TELEPHONE INTERVIEW

Physicians find the telephone interview challenging for many reasons
(Table 2-1). These challenges call for a more thorough and thoughtful

---

Table 2-1   Challenges in the Telephone Interview

- Inability to perform a physical examination or office-based tests (e.g., urinalysis/electrocardiogram)
- Altered communication patterns—physician is often in non-medical setting (e.g., home, restaurant)
- Occurrence during stressful situations (e.g., between office visits, after office hours, driving while using a cellular phone)
- Difficulty of documenting a call when in a non-medical setting
- Perception of calls as intrusions into the private life of the health care professional
- Brevity of calls compared with office encounters
- The following challenges are unique to the "on-call" physician taking calls for other physicians:
  – Little or no knowledge of the patient and no access to the patient's medical record
  – Lack of a pre-existing doctor-patient relationship

---

approach to the telephone interview. This chapter's aim is to give the reader an overview of the goals, structure, and functions of the telephone interview.

GOALS OF THE TELEPHONE INTERVIEW

Patient Goals

Patients who call often have different needs than they do when they visit in person. Patients often use the telephone as a convenient and economical first step into the health care system to "check things out" or to get help from a nurse or physician to organize or explain their symptoms into a comprehensible pattern. They can then decide, in collaboration with the clinician, whether to take some action, be it symptomatic remedies or arranging an office or a hospital visit. Stewart et al have described a taxonomy of illness behavior that provides some understanding of what pushes an individual to contact the physician (3). Calls are usually made because patients or their advocates have reached their tolerance limits for discomfort or anxiety.

When these limits are reached, they feel the need to obtain professional advice and reassurance. The physician should be aware that he or she will have to manage the caller's anxiety as well as any physical symptoms, and that, for many patients, management of anxiety may be the only physician intervention necessary.

Physician Goals

The physician's goals for the telephone encounter are different from the goals for the in-person encounter. The focus in the telephone interview is on determining if, when, and where the patient should be seen, providing symptomatic relief for patients who do not require in-person evaluation, and offering reassurance to the patient when appropriate.

THE THREE-FUNCTION MODEL OF THE TELEPHONE INTERVIEW: OVERVIEW

Although the goals of the telephone interview are often quite different from those of the in-person encounter, the key elements of the encounter are the same. The three-function model of the medical interview described by Lazare, Putnam, and Lipkin provides a useful structure on which to hang the core issues of the telephone consultation (4).

The three functions performed by the clinician are:

1. Determining the nature of the problem
2. Developing the therapeutic relationship
3. Educating the patient and implementing the treatment plans

Let us see how this model can be used to analyze, in the following example, Dr. Alexander's short telephone encounter with a patient suffering from low back pain.

CASE 2-1 **"Codeine worked for my back last time. I'm going to be travelling and I'm afraid to be without anything for the pain."**

Dr. Alexander is just about to leave his office at 7 p.m. after a long afternoon of patients. He is due to play racquetball with a friend at 7:30. His nurse catches him on the way out, saying that Mr. Walters, who happens to be the owner of the big furniture store in town, is on the phone with a recurrence of his back pain, and it is important that he talk with the doctor. Feeling irritated, Dr. Alexander picks up the receiver and says, "Hello, Dr. Alexander here." Mr. Walters responds cheerfully, "Hi, Doc!" (over the phone, he doesn't sound to be in much pain). "Sorry to bother you, but my back has gone out again—remember I saw you about 3 months ago because of my old back trouble—and you gave me some codeine pills for a couple of days, which helped a lot."

Dr. Alexander doesn't have the chart, can't remember this patient, and feels increasing irritation at the mention of narcotics and at the way his patient is calling him "Doc." "Oh, yes," he says in a neutral tone.

"Well," Mr. Walters replies, "I have to go out of town on business tomorrow, and I'm really stiff, and in pain. I know what exercises to do, and I use ice packs, but I am afraid to be without anything for the pain while I'm travelling. Can you help me, Doc?"

"How long has your back been giving you trouble?" asks Dr. Alexander.

"Oh, about a week. I was hoping it would get better, as it usually does, but it hasn't."

"Where's the pain located?" asks Dr. Alexander.

"Usual place," says Mr. Walters, "a couple of inches above the tailbone."

"Do you have any other symptoms? Any pain down the legs, any tingling or numbness?"

"No," Mr. Walters replies.

"Very well," says Dr. Alexander, knowing that he has a very busy clinic next day. "Come by tomorrow and pick up a prescription. I'll give you 15 codeine pills. That's one pill three times daily if needed, for 5 days. And please be sure to check with me when you get back from your trip."

"Sure, Doc," says Mr. Walters enthusiastically. "Anyway, I would really like to find out what you think about some new treatments for back pain, y'know, those magnets everyone is talking about. I'll be round first thing tomorrow to pick up the prescription. Thanks!"

"No problem," says Dr. Alexander, who has no idea about the use of magnets in low back pain. He puts down the receiver, picks up his sports bag, and leaves the office.

This case illustrates several positive and negative physician responses and some common aspects of the telephone interview. The three-function model can be used as a structure to analyze these aspects.

Function 1: Determining the Nature of the Problem

The clinician should take an effective history (which gives the diagnosis in 60%-80% of cases) (4), evaluate the patient's response to his or her illness or problem, and generate and test hypotheses.

In the above case, the patient, Mr. Walters, recognized the recurrent pattern of nonspecific mechanical low back pain and offered his diagnosis to Dr. Alexander, who appeared to go along with it. Dr. Alexander tested this hypothesis briefly with a screening question for possible neurologic compromise from a nerve root impingement.

An essential challenge of Function 1 on the telephone lies in determining the patient's primary concern. In this case, the primary concern is probably a combination of the common fear of unrelieved or persistent pain and anxiety about being unable to function effectively while travelling and working out of town.

However, Dr. Alexander spent little time acquiring psychosocial data and exploring Mr. Walter's agenda; he was, after all, in a hurry and probably made a decision not to deal with that issue.

Function 2: Developing the Therapeutic Relationship

The physician should demonstrate expertise, communicate respect and support, and address communication.

Mr. Walters seems to know Dr. Alexander fairly well, but Dr. Alexander could not recall his patient. This is a common occurrence in practice: a medical record is often unavailable, and there is no familiar face to trigger a physician's memory. Dr. Alexander, however, made little effort to build rapport. He did not respond verbally to the cues of psychosocial stress given by the patient, perhaps because he did not notice them. Despite the lack of any empathetic statement by the physician, the patient appeared satisfied because he got what he wanted—pain medications and an opportunity to discuss, at a future date, new treatments of low back pain. However, the therapeutic relationship is always a two-way street, and Dr. Alexander, irritated by this call, may not have performed as well as he would have if he had seen the caller in a more positive light. Additionally, this phone call may affect how Dr. Alexander perceives Mr. Walters at the next office visit.

Function 3: Educating the Patient and Implementing the Treatment Plan

The clinician should help the patient understand the significance of the problem or illness, recommend appropriate action and treatment, and clarify the actions and side-effects of medications and procedures. An essential task is for the physician to obtain the patient's agreement about the treatment plan, while taking into account both the medical and the psychosocial issues related to the patient's problem.

Much of this management plan, however, came from the patient, Mr. Walters: the request for a specific narcotic and an appointment to discuss a new type of therapy. Dr. Alexander agreed to the patient's management plan, although he appropriately proposed a follow-up arrangement to review the issue of narcotic therapy, recheck physical findings, and discuss magnet therapy. Dr. Alexander did discuss a narcotic dosage schedule and also required that the patient come by the office to pick up the prescription, thereby providing himself an opportunity to check Mr. Walter's medical record more thoroughly. However, he could have advised the patient on techniques for managing a stiff painful back while travelling in an automobile or by plane and mentioned adverse effects of the medication, including the need to avoid the use of alcohol in combination with the narcotic medication.

After reviewing the tasks of the three-function model of the interview, the reader might think that Dr. Alexander performed rather poorly in dealing with Mr. Walters on the telephone. However, in real-life practice, many circumstances, such as those listed in Table 2-1, prevent the ideal interaction from taking place.

Given Mr. Walters' demanding and somewhat aggressive behavior, the timing of the call at the end of the day, and the deadline imposed by the patient, perhaps Dr. Alexander did not do so badly after all. The quality of telephone care in this case was related to several immediate issues, such as those listed in Table 2-2, and although there was clearly room for improvement, Dr. Alexander did address some essential issues.

Table 2-2   Issues Relevant to Quality of Care in Case 2-1

Issue	*Implications for Clinical Care*
Prescribing narcotics	Useful for pain relief and patient satisfaction; needs monitoring
"Giving in" to patient demands	Maintained therapeutic relationship
Not taking a detailed history	Obtained just enough data to exclude some serious diseases but could have been more thorough (e.g., past medical history might include history of cancer)
Not demonstrating empathy for patient's distress	Could be addressed later but would have helped cement the therapeutic relationship
Lack of advice on traveling with back pain	Omission may lead to increased pain but could be mentioned when patient comes in for prescription
Documentation of call	Essential but unknown whether this was done
Follow-up	Negotiated satisfactorily with patient

THE TELEPHONE INTERVIEW AND THE THREE-FUNCTION MODEL: SKILLS AND APPROACHES

The above case and discussion provide an introduction to the three-function model as it applies to the telephone interview. Mr. Walter's case illustrates the complexity and the pitfalls of trying to practice good medical care over the phone. To effectively perform the three functions over the phone, the physician needs to maximize his or her skills in communication, both listening and speaking. The following discussion addresses the three functions individually and recommends specific skills and approaches that the physician can use to improve the practice of telephone medicine.

Function 1: Determining the Nature of the Problem

CASE 2-2   "I've been having this stiff neck all day and am very worried about it. I think I need to go to the emergency department."

At 11:20 on a Saturday night, Dr. Perez receives a call from Joe Bennett, a 38-year-old man who usually sees another physician in Dr. Perez' coverage group.

Dr. Perez: "This is Dr. Perez. How can I help you?"

Mr. Bennett: "Hi Doctor, this is Mr. Bennett. I've been having a stiff neck since this morning."

Dr. Perez: "Tell me about your stiff neck."

Mr. Bennett: "Well, it hurts in the back of my neck, especially when I turn my head. I woke up with it."

Dr. Perez: "Uh-huh...."

Mr. Bennett: "It's not too bad...I took some aspirin when it started, which helped a bit...but it's been going on all day. Do you think I need to go to the emergency department or something?"

Dr. Perez: "It sounds like you're pretty worried about it. I'd like to ask you some more questions to help me tell if you need to go to the emergency department. Is that OK?"

Mr. Bennett: "Sure."

Dr. Perez then asks some specific questions, which support his suspicion that this stiff neck is likely to be muscular in etiology, and not an emergency. While asking these questions, he notes that Mr. Bennett seems unusually worried. He wonders why the patient has called at all, let alone why he wants to know if he needs to go to the emergency department. He decides to explore this.

Dr. Perez: "I think it's very unlikely that your stiff neck is caused by something dangerous. Is there something specific you're worried about?"

Mr. Bennett: "Well, you're going to think I'm crazy, but I just saw on the 11 o'clock news that there's been an outbreak of meningitis at my kid's school. My kid feels fine, but now I'm worried."

This case demonstrates a variety of skills in performing the first function of the telephone interview. Dr. Perez begins with a positive introduction and an open-ended question ("This is Dr. Perez. How can I help you?"), encourages the patient to describe the symptoms fully ("Tell me more about your stiff neck"), listens without interruption ("Uh-huh...."), acknowledges the patient's concern and question about emergency evaluation without giving a premature answer ("It sounds like you're pretty worried about it. I'd like to ask you some more questions...."), then fully explores the symptoms. Dr. Perez picks up on the patient's seemingly inappropriate degree of worry and asks about it ("Is there something specific you're worried about?"), and uncovers Mr. Bennett's heretofore hidden concern about meningitis. Without this information, Dr. Perez would likely have been unable to satisfactorily reassure Mr. Bennett. This case demonstrates that this first function of determining the nature of the problem must be fully performed in order to provide effective care.

Challenges in Determining the Nature of the Problem

Several telephone studies have shown inadequacies in history taking by physicians (5,6). In a study comparing telephone management skills of pediatricians, it was found that nurse practitioners consistently obtained more historical information than house officers or practicing pediatricians (5).

Although no studies have been done that demonstrate that inadequate history taking leads to poor outcomes, Levy showed that in calls to an emergency room setting, nurses and physicians often failed to obtain essential baseline information. A study of calls to an emergency department concerning children, for example, found that in more than half of the calls the child's age was not elicited (7).

On the telephone, the medical history does not often unfold in a rational sequence. How a patient presents his or her symptoms, and what he or she leaves out of the story, are both important issues. As clues or items of information arise in the conversation, the physician may have to backtrack at times or reassess his or her initial diagnostic hypothesis.

Effective data gathering on the telephone involves being thorough—as thorough as in an office visit—and using effective communication skills and techniques to foster open and complete responses from patients. These two tasks are related, as Case 2-2 illustrates. Table 2-3 summarizes the essential elements of history taking on the telephone and the communication skills required to accomplish these tasks.

Tasks and Skills in History Taking on the Telephone

The following discussion elaborates on the elements of history taking and necessary skills as summarized in Table 2-3.

Opening
A welcoming introduction sets the scene for a good interaction and immediately gives the patient the feeling that the physician cares and is interested in helping.

Table 2-3   Essential Elements of History-Taking and Communication Skills

Essential Element of History	*Communication Skill*	*Typical Question/Statement*
• Opening: physician introduces self and obtains name, phone number, age, primary physician's name	• Use of positive opening; proper clarity, rate, and tone of speech; appropriate tone of voice; obtain basic demographic information	• *Hello, this is Dr. X; how may I help you?* • *Before we get started, let me make sure I have your full name and telephone number.*
• Identify primary concern(s)/actual reason for calling (ARC)	• Open-ended questions • Active listening for underlying concerns	• *How may I help you?* • *Is there anything else you need help with right now?* • *What's bothering you the most right now?* • *Is there something specific you want me to do for you?*
• Organize flow of call into a clear narrative	• Begin with open-ended questions • Encourage patient to elaborate without interrupting • Clarify unclear descriptions • Focus on the main problem • Use clear specific questions to fill in the gap • Remain tuned for new data/verbal cues hinting at ARC	• *How did this start?...Then what happened?* • *Uh-huh...Tell me more about that.* • *What do you mean by that?* • *I think we should focus on your headaches for now; it seems as though that's your main concern. Is that OK?* • *Are you having any difficulty breathing?*
• Ensure the description of primary concern is complete	• Repeat back the patient's history • Ask close-ended/specific questions to "fill in the blanks"	• *Let me repeat back to you what you've told me.* • *What have you tried so far to help the headache?*
• Gather relevant background information: comorbid conditions, medication, allergies, psychosocial history, social and/or family history	• Begin with open-ended questions • Ask clear simple questions to fill in gaps	• *Do you have any medical problems?* • *Can you tell me how things are going at home?* • *Do you have any allergies to medications?*
• Assess patient's beliefs about problem, sense of urgency	• Listen actively • Ask about the patient's symptoms	• *What do you think is happening?* • *Are you worried that something bad is going to happen?*

INTRODUCE YOURSELF CLEARLY WITH A STATEMENT OF CONCERN

It is polite and useful to give your name. An opening such as "This is Dr. Perez. How can I help you?" is welcoming and encouraging. Questions such as "What's the problem?" or "What's the matter?" can be interpreted by patients as hostile or abrupt, possibly leading to a reluctance to disclose personal information.

USE PROPER CLARITY AND RATE OF SPEECH

The clarity and rate of speech can suggest impatience, inattentiveness, or concern to the listener. On the phone, it is best to speak a little more slowly, especially if talking with elderly patients (who may have undiagnosed hearing deficits) or those from different ethnic backgrounds (who may have difficulty with language/accents).

USE AN APPROPRIATE TONE OF VOICE

Because visual communication of empathy and interest is not possible, variation in tone and warmth is vital to the communication process. The physician's voice naturally may suggest empathy, interest, boredom, sympathy, or arrogance, so it may be useful to develop a particular telephone voice style. The physician's family members may be able to give excellent advice to the physician on his or her usual voice style on the telephone and how it could be improved.

OBTAIN BASIC DEMOGRAPHIC INFORMATION

It is essential to record the patient's name and age, the time and date of the call, and the patient's telephone number and his or her personal physician's name. It is useful to know where the patient is calling from because his or her location may affect decision making and management. Additionally, the physician may need to call back for either clinical or administrative reasons, or simply because of disconnection of the line. Basic demographic information is often recorded by a nurse, receptionist, switchboard operator, or answering service before the physician speaks to the caller, but the physician should request and record this information as well.

Identify the Primary Concern(s)/Actual Reason for Calling

Just as in the office visit, the patient's real agenda may be quite obvious or may be hidden (consciously or subconsciously) behind an offering of symptoms or concerns. Thus the expression "My stomach is aching" could mean true pain or represent stress or some dependency need. The patient's symptoms may not truly have a biological cause. Even when they do, the actual reason for calling (ARC) may be more related to anxiety about the symptoms rather than the discomfort. In Case 2-2, Mr. Bennett's neck pain was not that bothersome; it was his concern about meningitis that made him uncomfortable and prompted him to call.

The ARC is usually going to be that the patient (or companion) has reached his or her tolerance limits for discomfort (e.g., pain, bleeding, dizziness, etc.) or anxiety (concern about the implications of their symptoms).

When these limits are reached, the patient feels the need to obtain professional advice and reassurance (3). As seen in Case 2-2, knowing the ARC is instrumental in helping the caller. Dr. Perez needed to realize that Mr. Bennett's anxiety had to be managed, perhaps even more than his physical symptoms. This is likely to be true for many callers, who may be looking for reassurance rather than medication. Indeed, one study found that 66% of callers felt that reassurance, explanation, and advice were the most valuable aspects of the consultation (2). If the physician can discover the patient's ARC early in the call, he or she can address it more fully and efficiently.

USE OPEN-ENDED QUESTIONS

As in the office setting, questions can range from the general to the specific. An opening statement by the physician (such as "What can I do for you?") will lead to the broadest range of information; therefore, the physician should allow the caller some time to tell his story without interruption.

ASK ABOUT OTHER/UNDERLYING CONCERNS

When the patient finishes his or her description of the problem, the physician should check with the patient about other concerns, which may be related to or completely different from the first problem reported. A question such as "Is there anything else you need help with right now?" may uncover other important concerns, perhaps even more important than the first. When all these concerns are "on the table," the physician and patient can start to talk about the most important concern with a question such as "What's bothering you the most right now?" When the ARC still remains unclear, a question such as "Is there something specific you want me to do for you now?" or "Is there anything else going on in your life that you want to tell me about?" Open-ended questions such as these will increase the likelihood that the patient brings up the important issues early in the interview, and may save time spent pursuing issues of lesser importance to the patient.

It may not always be possible to determine the ARC right away. Sometimes patients may need to "test the waters" a bit before feeling comfortable with the physician. Other times, the patient will not be able to articulate the ARC at first. For these reasons, physicians should stay open throughout the duration of the call for cues that suggest the ARC is different from what the physician presumed it was.

Organize the Flow of the Call into a Clear Narrative

Have the caller describe the problem chronologically and in general terms and then move to specific questions. This will allow for a more complete, logical, and efficient description of the problem. Be prepared to backtrack if new clinical or behavioral cues appear.

Patients who ramble should be managed as recommended in Chapter 17. Some callers may speak in generalities or abstractions and be reluctant to be pinned down by specific questions, which may suggest a hidden agenda or an emotional problem.

ENCOURAGE THE PATIENT TO ELABORATE WITHOUT INTERRUPTING

In person, a physician's open posture and attentive silence may encourage patients to express themselves. On the phone, patients may misinterpret silence as a lack of interest or understanding. Simple expressions such as "Uh-huh" or "Go on" or "Tell me more about that" convey the physician's interest and encourage elaboration without breaking the patient's line of thought.

Sometimes the physician will be silent for an extended period because he or she is performing another task. Simple statements such as "Please excuse me a moment, I am just checking your medical record" or "Just a minute—I need to think more carefully about your problem [or look something up]" will avoid misinterpretation on the patient's part.

CLARIFY UNCLEAR DESCRIPTIONS

Given social and cultural differences and other possibilities for misinterpretation, the physician must know when to use clarifying questions (i.e.,

Patient: "Doc, I've been going to the toilet all night." Doctor: "What do you mean by that?" or "Was it your bowels or a problem with your urine?").

The physician's definition of a word maybe quite different from the patient's, especially if the patient and physician are from different cultures. For some ethnic groups from the Far East, headache may represent mental distress, whereas in some Latin American cultures, a word for fatigue may also mean depression or even dyspnea. Terms for anatomic locations may have different meanings: one patient's "stomach" may be just above the pubic bone, while another's is over the liver. Physicians should use less ambiguous anatomical landmarks to clarify the location, such as "Is the pain above or below the belly button?"

Another possible source of miscommunication is hyperbole. "I've got this excruciating, horrible pain in my leg" needs to be checked out by clarifying questions such as "How far can you walk?" The physician needs to determine if the patient is highly anxious and using this style to gain attention or really has a major problem.

FOCUS ON THE MAIN PROBLEM

Patients who are nervous may be distracted by less important issues. To ensure adequate attention to the primary concern, physicians should say something such as "I think we should focus on your headaches for now— it seems like they're your main concern. Is that OK?"

The opportunity for a patient to speak directly with his or her physician in the comfort of his or her home can lead to many "and by the way" issues, such as discussing recent lipid tests, the spouse's poor sleeping habits, and problems communicating with an HMO! The physician can gently but insistently help the patient focus by using redirecting statements such as "Just a minute, if it's okay with you, I would like to spend more time discussing the swollen leg you called about?" or, more strongly, "I'd like to go back to the main reason you called today. I don't think we have dealt with it fully yet."

REMAIN "TUNED-IN" FOR NEW DATA/VERBAL CUES HINTING AT THE ARC

On the phone, physicians may jump to a diagnostic conclusion early in the consultation and then exclude new information that does not fit their initial diagnostic thinking (8). Sometimes the caller may present new cues later on in the interaction that may indicate the existence of another problem (e.g., "Well, my asthma has been especially bad at work," may be a hint about a stressful work situation or a potential environmental allergen). Patient fears are often expressed as questions starting with "What if?" or "Could it be?" (e.g., "What if my asthma gets worse?").

In Case 2-2, Mr. Bennett's question about a emergency evaluation for what seemed initially to be a routine stiff neck suggested to Dr. Perez that the ARC was the patient's worry and not his discomfort. Again, it is not always possible to determine the ARC right away. As in Case 2-2, the patient may have to discuss the "chief complaint" a bit before feeling comfortable enough to tell the physician about what is really the primary concern. When the physician suspects the ARC is something other than the chief complaint, useful questions include "Is there anything else going on that you want to tell me about? and "How would you like me to help you now?"

During the telephone consultation, new data will often be heard that do not correlate with the initial diagnosis. For example, when a physician identifies the problem as mild gastroenteritis and begins to recommend clear fluids with reassurance that recovery is soon to be expected, the patient may unexpectedly throw in new information, such as "What about the rash I've got on my legs?" Such information, added after the physician has made his or her diagnosis, can often be discounted or minimized by the physician, because it does not seem to fit. The physician must always be prepared to rethink the differential diagnosis and the management plan based on new information.

Patient self-diagnosis can be a pitfall on the phone. The physician must be careful not to be too accepting or misled by the patient's explanation of the problem. In Case 2-1, Mr. Walter's offered his own self-diagnosis of "my old back trouble," but Dr. Alexander nonetheless asked about symptoms of nerve impingement. A caller's certainty that he has "sinusitis" or that she has "menstrual cramps" may mask a more threatening medical situation that the physician fails to uncover because he or she accepted the caller's assessment of the problem.

Make Sure the Description of the Primary Concern is Complete
Given the sometimes rushed circumstances on the telephone, it can be difficult to ensure that the history taken is complete. Table 2-4 provides an organizing framework for describing the primary problem, including information that should be obtained in any telephone consultation (9). The

Table 2-4 Framework for Describing the Primary Problem

Component	Typical Questions
Chronology	*When did it begin?* *Is it getting better? Worse?*
Anatomic location	*Where exactly is the pain?* *Is it above or below your belly button?*
Quality of the symptom	*Describe the pain for me.* *Is it like a weight on your chest, or is it more like a knife?*
Quantity	*On a scale of 1 to 10, with 10 being the worst pain you've ever had, how bad is this?* *How often have you been having bowel movements?*
Aggravating and alleviating factors	*Does anything you do make it worse?* *Have you tried anything to relieve the pain?*
Associated symptoms	*Have you noticed anything else new?*

shorter time spent on telephone consultations likely results in a tendency to overlook some of these components of the chief complaint, but no history is complete without them.

The following skills help to ensure that the physician's knowledge of the patient's symptoms is as complete as possible.

REPEAT BACK THE PATIENT'S HISTORY

This simple task ensures a complete history in a number of ways. First, it gives the physician a chance to go over the facts in his or her head. This provides a chance to recognize inconsistencies or missing information that can be clarified with the patient. Second, it gives the patient a chance to hear what the physician's understanding of the facts is, allowing the patient a chance to correct any misunderstandings, add new information, and be assured that the physician has been taking the problem seriously.

ASK CLOSE-ENDED/SPECIFIC QUESTIONS TO "FILL IN THE BLANKS"

After gathering the history in an open-ended manner as described above, the physician will often need some specific information not spontaneously offered by the patient. These gaps can be filled in using simple, clear questions, such as "Is there any blood in your bowel movements?" Leading questions (e.g., "Do you have a sharp pain in your right side?" or "You don't have any fever, do you?") should be avoided because the style of questioning may alter the patient's response. Avoid multiple questions such as "Where exactly and how long have you had this swelling, and how often does it come on?" Asking the questions individually will allow for more complete responses.

When a patient is unable to respond to open-ended questions such as "How would you describe the chest pain?", more closed-ended questions are necessary. In this case, the question should be asked in a non-leading fashion, such as "Would you say the pain is like a weight on your chest, or is it more like a knife?" This will allow a specific response without the patient saying "yes" just to convey that the pain is severe.

Gather Relevant Background Information

ASK ABOUT MEDICINE HISTORY AND CO MORBID CONDITIONS

Other elements of the medicine history may or may not be relevant, depending on the type of problem as defined above. Patients may not volunteer this information, perhaps because they are concentrating on the "new" problem. Nonetheless, physicians may neglect to ask about co-morbid conditions and medications (prescription and non-prescription). Another surprising and not infrequent aspect of the history that is omitted is asking about other medical problems or medications.

TAKE A PSYCHOSOCIAL HISTORY

Is the caller under particular stress because he or she is unemployed or a student taking final exams? When it appears obvious that the call is made because of mental health problems or because of stress and social problems, additional data are needed. These should include the patient's current occupation, the patient's living situations, details of his or her recent stressors, and social support systems. The sexual history is often helpful but is often avoided over the phone. Assuring the patient of privacy and explaining why a sexual history is important will likely make the patient more comfortable providing it.

INQUIRE ABOUT SOCIAL AND FAMILY HISTORY

Social and family history may or may not be relevant in a telephone encounter. The decision to see a patient may be partly based on the availability of family or friends either to bring the patient in or to care for him or her until the next day. Obtaining a social history on the telephone may be crucial when there is a possibility of domestic violence, alcoholism, depression, or drug ingestion. Social and family histories generally will be relevant in calls with hidden agendas. In other calls, such as those for acute trauma, chest pain, and other emergencies, it is not necessary to ask any social history questions other than "Do you have a way of reaching the hospital, clinic, or office?"

Assess Patient's Beliefs About the Problem and Sense of Urgency
It is often worthwhile to ask a patient "What do you think is causing your symptoms?" The patient may come up with a plausible diagnosis that the physician had not considered. More commonly, the patient's response will guide the physician in effectively reassuring the patient.

Similarly, it is important to get an idea of how serious or important the caller thinks the problem is. A statement such as "Are you worried that this might get worse quickly?" is helpful in getting at the patient's sense of urgency. The physician can then correlate this information into assessment and management of the problem. Often, however, if one asks directly, "How urgent do you think this problem is?" the caller may feel pressured by the context of the call to confirm the urgency of the problem and may become somewhat defensive.

Common Barriers to Data Gathering

CASE 2-3 A Third Person Calling on Behalf of the Patient

Dr. Evans receives an urgent call from the home of Mary Firth, a 60-year-old woman with diabetes and hypertension. The caller, her neighbor, breathlessly says "Mary has collapsed in the kitchen. Shall I call the rescue squad?" Dr. Evans pauses and then asks, "Is she conscious?" "Oh, yes," answers the neighbor. "Well, can you tell me what she is doing now?" asks Dr. Evans. "She's sitting in a chair drinking a cup of tea," the neighbor says. "Do you think she is well enough to talk to me?" says Dr. Evans. "Oh, I think so," says the neighbor, "I will put her on the line."

Table 2-5   Common Barriers to Data Gathering on the Phone

- Technical problems with the phone connection/equipment
- Hearing/speech deficits
- History transmitted through family member
- Privacy concerns (especially with cellular phone)
- Language barriers (including dialects)
- Cultural differences (in communication styles, disease models)
- Conflicting/simultaneous physician activities
- Patient hostility/unwillingness to answer questions

There are many barriers to data gathering on the telephone. Table 2-5 lists some of the more common ones. Some barriers are obvious: technical problems, hearing/speech deficits, needlessly talking to the patient through a family member, and language differences all should be addressed immediately. Technical problems can sometimes be solved by trying a different phone, and family members should only be involved when the patient has a hearing, speech, or language barrier. When the physician is busy and distracted by other tasks, and the patient's problem is not an emergency, the physician can ask the patient's permission to call back at a set time later that day.

Barriers arising from differences in culture or communication style may be harder to recognize and address. Many studies have shown that differences in cultural views can lead to a failure to communicate or to understand a problem. Often, doctor and patient do not respond to each other because one does not understand the other's words or the social context of the other's beliefs and actions (10). For example, certain patients are not assertive and expect the clinician to drag a history out of them. An open-ended questioning style (e.g., "Why don't you tell me what's going on?") may be met by a long silence or seemingly illogical speech at the other end of the line; therefore, very specific closed-ended questions must be used.

In Case 2-3, Dr. Evans could have easily fallen into the trap of assuming it was the neighbor calling because Ms. Firth was unable to speak. By directly questioning the patient, the history will be more complete and reliable. Barriers to communication such as this are obvious impediments to acquiring accurate information. In this case, Ms. Firth's "collapse" may have been no more than a slip on a wet floor with no clear injury. Often the physician needs to do little more than recognize the barrier, and removing it may be easy.

The physician must also be prepared to deal with difficult or frustrating communication issues, such as reluctance to disclose information. For example, on being asked for more information, the patient might say in an annoyed tone, "I don't see what your questions have to do with what I'm calling about!" This type of comment will often kindle a flame of resentment and frustration in the physician, which must quickly be extinguished, or kept at a low level until the end of the call (11). Realizing that the patient's comment may flow from acute anxiety should allow the clinician to reassure the patient while still asking the appropriate questions to rule out serious pathology. If there is overt hostility from the caller, it is better to confront the situation with statements such as, "You sound dissatisfied with what I just said" or "You seem very angry about the situation" (see Function 2, below, and Chapter 17).

Function 2: Creating Rapport and Building a Relationship with the Patient

On the telephone, rapport and relationship building can be especially challenging, but they can also be vital to a successful telephone interview. For a patient to be reassured that his stiff neck is not life threatening, for example, requires that the doctor develop the patient's trust, respect, and confidence. Similarly, the doctor's trust in the patient affects his confidence that the patient will carry out the treatment plan.

CASE 2-4 Is It My Heart?

A 32-year-old postal worker, Andy Reef, calls the office to speak to his physician about chest pain. According to reception desk protocol, he is put directly through to Dr. Miles. On taking a history, Dr. Miles notes that the pain is sharp and burning in nature, intermittent, located up and down the sternum, and not made worse by effort, but increases on lying down and after eating. Dr. Miles has treated the patient for gastroesophageal reflux in the past and suggests that Andy restart his previous medication regimen.

Dr. Miles notices that Andy responds in a very hesitant way, as though he was somewhat distracted or tense and reluctant to take the medicine.

Dr. Miles: "Andy, you don't sound very comfortable with my suggestion. Is anything else going on?"

Andy: "Well, my dad, who lives in Florida, was just admitted 3 days ago with a heart attack, and I seem to have the same symptoms."

Dr. Miles: "I see. How is your father doing?"

Andy: "Well, it sounds pretty bad. He's in intensive care. I was hoping to go down there tomorrow, but now I'm worried if maybe I need to get checked out myself."

Dr. Miles: "It sounds like you've been pretty worried about your father, and now you have symptoms that you're worried about as well."

Andy: "Yeah, that's right. What should I do?"

Dr. Miles: "Whenever anyone tells me they're having chest pain, I have to consider the possibility that it may be a heart problem. You've told me that your pain is sharp and burning, that it gets worse with eating and lying down, and that it doesn't get worse with exercise like going up stairs. Does that sound right?"

Andy: "Yes, that sounds right."

Dr. Miles: "I'm pretty confident that this pain you're having is not from your heart. Heart pain, what we call "angina," is usually a pressure-type pain, and is usually worse with exercise. Food doesn't usually affect it. It sounds very much like you're having an episode of reflux again. Does that make sense?"

Andy: "Yes, I think it does. I guess I just wanted to be careful."

Dr. Miles: "I'm glad you were. You've always taken good care of yourself, and being careful is part of that."

Andy: "So should I start back on my reflux medicine?"

Dr. Miles: "I think that's a good idea. But call me back right away if the symptoms change, or if the medicine doesn't help, or if you have any questions."

Andy: "I will. Thanks a lot, Dr. Miles."

Tasks and Skills in Building and Maintaining Rapport

A good therapeutic rapport is vital to effectively managing cases such as Mr. Reef's. Mr. Reef must feel comfortable before disclosing his underlying fears to Dr. Miles, and Dr. Miles must instill in Mr. Reef confidence that the problem has been thoroughly assessed and that the recommendations are sound.

The main tasks involved in building the rapport are:

1. Recognizing behavior and language that signal an underlying problem
2. Assessing and managing the patient's emotional state and other underlying problems
3. Instilling confidence in the patient

Each of these tasks requires specific skills to ensure their effective completion.

Recognizing Behavior and Language That Signal an Underlying Problem

As discussed above in Function 1, the patient's underlying concern, the ARC, may not be expressed readily on the telephone, especially if the patient does not know the physician. Uncovering an underlying concern, especially when it involves strong emotions, is essential to developing a therapeutic rapport. Often the ARC is fear that the presenting symptoms represent a deadly condition such as cancer, or that they are similar to a relative's serious illness. This may be the case even in healthy young adults. The ARC is not always fear of a bad medical outcome. Table 2-6 lists some examples of the wide variety of underlying concerns that may be the ARC.

REMAIN OPEN TO NEW CONCERNS THROUGHOUT THE CALL
This is not always easy. Often, after the physician feels he or she has "done his or her job" (identified and managed the problem), the patient signals

---

Table 2-6 Categories and Examples of Actual Reasons for Calling

- Administrative (frustration with the office appointment system)
- Medical (fear of cancer, bleeding, infection, life-threatening illness)
- Psychological (depression, borderline personality, drug dependence)
- Social (family or marital dysfunction)
- Occupational (fear of inability to work, desire to avoid work)

that an underlying concern exists. Sometimes this can be identified "up front" (see Function 1), but often the patient may not disclose the ARC until much later in the call (e.g., after a treatment plan has been recommended), as Mr. Reef did in Case 2-4.

IDENTIFY SIGNAL BEHAVIOR/LANGUAGE

Physicians can often identify signal behavior or language that indicates that the problem the physician is addressing is not the ARC. The challenge is to recognize when language/behavior differs from what one would expect with this patient given the problem confronted. Paralanguage (non-word communications) can be as important as words in conveying meaning. The physician can use his or her awareness of nonverbal signals to gain valuable information about the patient. For example, the caller's rapid speech patterns (anxiety?), long pauses (apprehension? difficulty in comprehension?), high pitch (anxiety? hostility?), monotonous intonation (depression?), may lead the physician to inquire into the "hidden agenda" and psychological status of the caller (12). Table 2-7 lists some examples of signal behavior and language that may be signs of stress, emotion, or unmet need.

In some cases, the physician will perceive no "signals." Even when no signal behavior or language is present, physicians should still ask themselves if the initial complaint is the ARC. It is often useful for clinicians to ask themselves (or the patient), "Why is the patient calling now?" Asking

---

Table 2-7 Common Examples of Signal Behavior and Language

- Caller's voice tone and pitch
- Pressured speech or long pauses
- Delayed answers to questions
- Repeated words or phrases
- Dissatisfaction with physician's recommendations
- Patient's distress seems out of proportion to the problem
- Patient perceived as hostile, demanding
- Unreasonable or apparently trivial complaint
- Strong emotional response in physician

Table 2-8   Statements and Questions to Assess the Actual Reason for Calling

- *Are you okay?*
- *I'm getting the feeling that something is upsetting you.*
- *Is there anything else that concerns you?*
- *I'm wondering if there's something bothering you that you haven't told me about yet.*

this question may help to reveal underlying concerns. The physician should remember that the limits of a patient's tolerance are not related purely to severity of symptoms but to many personal and social factors as well.

Mr. Reef's signal behavior is his persistent worry despite Dr. Miles' seemingly appropriate reassurance. This is an indication that Dr. Miles has more work to do before he has completely and effectively managed Mr. Reef's problem.

Once the physician suspects that the ARC is something other than the problem(s) already being addressed, he or she must encourage the patient to discuss it fully. Then the physician can direct further efforts to addressing this concern. Here, the first and second functions of the interview can be served together because developing rapport allows the patient to communicate more openly, which in turn leads to further rapport building. Simple statements expressing concern, such as those in Table 2-8 can be effective in encouraging patients to open up.

These statements will often allow the patient to verbalize his or her distress. Once the physician is aware of the ARC, it can be managed (see discussion on Skills for Emotion Handling and Relationship Building, below).

Another approach to uncovering the ARC is to inquire about the patient's perspective. Statements such as "What do you think the bleeding might be caused by?" or "Are you worried that this might be serious?" will help to identify patient beliefs and fears that will need to be addressed.

Summarizing the patient's history back to him or her may also allow the patient an opportunity to add in the ARC, especially after being reassured that the physician has been listening attentively.

Assessing and Managing the Emotional State
If the concern uncovered by the above approaches is medical, it needs to be evaluated fully as a possible medical illness. If the concern is emotional,

such as fear, the physician must use skills in emotion handling to ease the patient's suffering. The approach with other emotions, such as sadness or anger, is comparable (see Chapters 15 and 17).

Skills and behaviors that may be effective in emotion handling and relationship building with patients on the telephone such as Mr. Reef primarily focus on the communication of empathy. Simply conveying the message that the patient is being understood and cared about can in itself be very therapeutic (13). On the telephone, communicating to a patient that his or her concerns are appreciated and understood can be quite challenging. In person, a patient can see a concerned posture or facial expression, and the physician's touch can express deep understanding of the patient's situation. On the telephone, the physician has only words and the sound of his or her voice to communicate empathy. Table 2-9 lists specific skills that can be helpful in the verbal communication of concern and understanding (13).

The physician should use *reflective statements* to "reflect back" to the patient what emotions the physician has perceived. "It sounds like you're pretty worried about having a heart attack" or "It sounds like you've been waiting a long time and this bleeding is really making you upset" or "You sound very upset about this problem." Another effective use of a reflective statement would be with an angry patient. This can sometimes be addressed after dealing with the medical aspects of the problem, but if the caller cannot get past the anger, it should be addressed immediately or it may prevent effective communication. "Mr. Smith, you sound quite upset. Is there something I can do right now?" or "You sounded upset when we first spoke. Was there anything about this problem that I did not deal with?"

Legitimizing by telling the patient that his or her emotional response is normal and understandable can be comforting and it assures the patient that the doctor appreciates the patient's concerns. For example, the

---

Table 2-9   Specific Skills in Communication of Concern and Understanding

• Reflective statement	• Expression of partnership
• Legitimation	• Communication of respect
• Supportive statements	

physician could say, "I can understand your being upset and concerned. A heart attack is a very serious problem."

Physicians may feel that their actions "speak for themselves," but *supportive statements,* clear statements that the physician is there to help the patient, may be very reassuring. For example, "I want to help you be sure whether or not this is a heart attack" or "I'm here to help you any way I can."

The physician should *express his or her partnership* by offering to work together with the patient to meet the patient's needs. For example: "I've got some ideas about how to make your symptoms get better and make sure this isn't a heart attack. Why don't I explain what the options are and we'll see if we can come up with a plan together."

Communicating respect by acknowledging the patient's taking responsibility for his or her medical care may help the patient feel better about his or her role in the treatment plan. "You've been quite honest and open with me about your concerns. That takes some courage over the phone. I think the decision you've made to try some antacids and call me back in half an hour is very reasonable."

Instilling Confidence in the Patient

Patients at home calling for assistance with a medical problem are clearly looking for expert advice. A vital task in any medical interview is to communicate professional expertise and confidence to the patient (4). This is especially challenging in a brief telephone encounter. However, the implications of a patient having a low level of confidence in the physician are great. When a physician seeing a patient in a medical office advises emergency department evaluation, he or she will know quickly if the patient plans to go with the ambulance when they arrive. On the other hand, once the telephone is hung up, the physician may not know if the patient plans to follow through with such a recommendation (unless the physician calls back later). A patient that doesn't have confidence in the physician's opinion will be less likely to follow through with recommendations.

Because of the absence of nonverbal cues such as facial expression, gestures, and body posture, vocal communication (rate of speech, volume and pitch of voice, frequency of pauses in speech, voice intonation, articulation, and word stress) assumes greater importance. Paralanguage, discussed above as useful in conveying the patient's emotional state, may

Table 2-10   Behaviors Effective for Instilling Confidence

- Speak with a firm, clear, and calm voice
- Make a thorough assessment
- Ask and respond to the patient's concerns and questions
- Summarize the essence of the problem
- Give diagnostic assessments and therapeutic options clearly and succinctly
- State uncertainties without sidestepping or covering up
- Make statements that reflect one's previous experience and current awareness in dealing with similar problems

reveal various aspects of the physician's attitude to the caller, even when the words the physician has chosen do not (4). For example, when responding to a call, the physician may first say " Hello, Dr. Forsythe here," and then continue to respond to the patient in a clipped monotone. This communicates to the caller a general lack of interest and empathy, even though the management of the call may be reasonable. Table 2-10 lists behaviors that may be effective for instilling confidence over the telephone.

SPEAK WITH A FIRM, CLEAR, AND CALM VOICE

A firm, clear, and calm voice communicates a general attitude of confidence. Pauses between words may be interpreted, and misinterpreted, by the listener in various ways. Compare the following: "Well, um, uh, I don't think, um, its, um, likely to be too serious" and "Well, I don't think it's serious." Indecision and uncertainty are often conveyed by pauses and rushed or mumbled speech, and they may additionally give the impression of lack of concern.

MAKE A THOROUGH ASSESSMENT

A complete assessment demonstrates knowledge of and interest in the patient's past medical history, and a thorough exploration of all the patient's concerns helps to convey expertise (4).

ASK AND RESPOND TO THE PATIENT'S CONCERNS AND QUESTIONS

Attention to the patient's specific concerns will obviously give the patient a greater level of satisfaction with the care provided.

SUMMARIZE THE ESSENCE OF THE PROBLEM
Summarizing conveys good listening skills and the sense that the physician fully appreciates the problem.

EXPLAIN DIAGNOSTIC ASSESSMENTS AND THERAPEUTIC OPTIONS CLEARLY
AND SUCCINCTLY
See Function 3: Treatment Negotiation and Patient Education, below.

STATE UNCERTAINTIES WITHOUT SIDESTEPPING OR COVERING UP
Frankly stating uncertainties demonstrates the physician's honesty and commitment to the patient's care. "At the moment, I'm can't be quite sure what's going on. I think it would be safe if we watched your condition for a few hours and speak later. You can call me back if your symptoms change."

MAKE STATEMENTS THAT REFLECT PREVIOUS EXPERIENCE AND CURRENT
AWARENESS OF DEALING WITH THE PROBLEM
Obviously, one should not make statements about experience and knowledge that one does not have, but it is useful to communicate with the patient in this way, assuming one does have knowledge of and experience with the patient's problem. For example, "Yes, your problem is not unusual" or "My experience with this problem is that ..." or "This is a common side effect of this medication."

Common Barriers to Creating Rapport with the Patient

As in other medical encounters, the success of the telephone consultation depends upon the immediate circumstances, the characteristics and needs of the speaker, the perceptions and experiences of the listener, and the communication process itself (14). Barriers to developing a therapeutic rapport arise from patient characteristics, physician (or practice management) characteristics, and the interview communication process itself.

Table 2-11 lists some specific barriers in each of these domains. When these barriers arise, the physician must take steps to address them. If left unaddressed, these barriers can negatively impact the quality of the assessment and subsequent management.

Table 2-11   Barriers to Developing a Therapeutic Rapport

Barriers Arising from Patient Characteristics
- Strong emotions (fear of illness, anger, hostility, depression)
- Firmly held expectations about management (antibiotics, opioids)

Barriers Arising from Physician/Practice Characteristics
- Negative past experiences with the patient
- Suboptimal emotional state ("having a really bad day")
- Competing priorities (other patients, meetings, social/family time, driving in traffic)
- Clinical uncertainty about the patient's case
- Disorganized/inefficient office management of phone calls

Barriers Arising from Communication Process
- Patient disagrees with management plan
- Patient feels shamed, humiliated, or insulted by physician
- Patient perceives physician is not attentive, thoughtful, or caring
- Physician frustration with vague/confusing patient narrative

---

Barriers Arising from the Patient

CASE 2 5   "I Called to Speak to a Doctor over an Hour Ago!"

Dr. Gibbons receives a message from his nurse that Joe Symonds, 28 years old with a history of irritable bowel syndrome, has had diarrhea since the morning. Dr. Gibbons knows Mr. Symonds well because he is a frequent caller with a short fuse. Dr. Gibbons calls Mr. Symonds back when he has a break between patients. Mr. Symonds is clearly irritated, saying, "I called over an hour ago because I've got diarrhea!" Dr. Gibbons thinks to himself sarcastically, "He's had this countless times before, but now he's going to die because I didn't call him back fast enough."

The anger expressed by Mr. Symonds led Dr. Gibbons to get irritated and begin to dismiss the complaint as trivial and annoying. Anger and hostility frequently arise as a result of the caller or patient having to wait for what seems to them to be an unreasonably long time for the physician to call back. Usually this hostility is easily detected by the physician, and it is important to defuse it with an apology. This may even involve having to apologize for one's own health organization, or, for example, for a colleague who did not prescribe an adequate amount of medication for the patient. Until the hostility has been diffused, the data obtained during the call will likely be flawed and the outcome less than satisfactory. If, in fact, the physician believes that the patient is "out of line" (for example, the caller has been rude to the office staff), then it is best to confront the issue at the end of the call, after the patient's problem has been addressed.

When confronted with an angry patient, the physician may need to "manage" the anger first, and then move on to the medical problem (see Chapter 17). Demonstrating an understanding for the patient's anger is a first step (see Skills for Emotion Handling and Relationship Building, above). Next, a straightforward apology can be effective in helping the patient to let go of the anger and move on.

Luckily, Dr. Gibbons does not verbalize his irritation and has a chance to start over, giving more attention to the possibility that something serious may be going on. Indeed, Mr. Symonds' anger may be a signal that there is, perhaps, something more to this call than his usual irritable bowel syndrome. Before he can explore this possibility, however, Dr. Gibbons needs to dispel Mr. Symonds' anger quickly and effectively. Allowing the patient to "ventilate" his or her feelings and responding with empathetic statements, as described above, are often effective in allowing the patient to feel "heard" and ready to move on.

Resolution of Case 2-5

Dr. Gibbons: "It sounds like you're pretty upset [reflective statement]. I'm sorry to have kept you waiting [apology]. I know it must be pretty important for you to call [respect]. I'd like to help you any way I can [support]. Why don't you tell me about the problem and we'll see what we can do about it [partnership]."

Mr. Symonds: "Yeah, we better, because I think this is something bad. It started this morning...."

This approach can be useful in a wide variety of situations in which the interview's progress is impeded by strong feelings such as patient sadness or worry. Another common barrier to communication on the telephone is family members or other companions calling for the patient, as in Case 2-6.

CASE 2-6 "You're Not the Doctor I Wanted to Talk To!"

Josephine Bennedetti calls the answering service on Sunday morning looking to talk to Dr. Davis about her husband Jake. Jake, 28 years old, has some bleeding from hemorrhoids, and both wife and husband are concerned about it. Dr. Parker, who is on call in Dr. Davis' practice, calls back.

Dr. Parker: "Hello, this is Dr. Parker. I'm calling for Jake Bennedetti."

Mrs. Bennedetti: "Dr. Parker? They told me I could talk with Dr. Davis! He's our family doctor, and I'd really like to talk with him!"

Very quickly, three barriers to communication can be identified: the caller is not the patient; this doctor is not whom the caller expected; and the caller is angry. Hopefully, Dr. Parker will be sensitive to this anger, which may be partly caused by fear of the implications of the bleeding or by some other personal stress. The disappointment about not speaking with the family physician, Dr. Davis, only adds to this anger. Clearly, Dr. Parker will have to manage this anger, much in the way Dr. Gibbons did in Case 2-5.

A twist in this case is that the caller is not the one with the medical problem. It is fairly common for family members, particularly women, to call about their children, adolescents, and menfolk, who may be reluctant to seek health care, particularly if it deals with an embarrassing part of their body. For Dr. Parker to effectively evaluate the situation, he must first deal with Mrs. Bennedetti's anger, and then try to speak directly with Mr. Bennedetti.

Resolution of Case 2-6

Dr. Parker: "I'm sorry they told you that you could speak with Dr. Davis. He isn't available today. I'm on call for him. Is there something I can help you with?"

Mrs. Bennedetti: "It's my husband. His hemorrhoids are bleeding, but I guess he's just too busy to call for himself."

Dr. Parker: "I suppose Jake might be reluctant to talk about this. Why don't you put him on the line, and we will see if we can sort this problem out."

Mrs. Bennedetti retorts, "Well, I hope you can get more out of him than I can."

Not only is Mrs. Bennedetti angry that she wasn't speaking to Dr. Davis, it seems she's also angry with her husband. This is clearly a serious communication barrier that would likely impair Dr. Parker's ability to evaluate the condition. In other cases, the dynamic between the patient and the actual caller may be more subtle but no less detrimental to communication. It is therefore imperative that the physician speak to the person with the medical problem whenever possible. Here, Dr. Parker made a simple, direct request to put Mr. Bennedetti on the phone. This will allow Dr. Parker to acquire information before it gets interpreted and filtered through a third party.

Barriers Arising from Physician and Practice Characteristics

CASE 2-7 "I Lost My Prescriptions."

Dr. Williams has been having a bad day in the clinic, mainly because she is beset by family difficulties that are distracting her. She has already had an argument with her nurse, who justifiably stood up to her and answered back. She gets a call from Mr. Wallach, a 65-year-old retired electrician with hypertension. "Sorry to bother you, Doctor," says Mr. Wallach. "I saw you a couple of days ago, and, unfortunately, I lost my prescription for the blood pressure medicine. Could you call it in for me? I'm really sorry."

"I suppose so," says Dr. Williams, in angry tone. "You really need to pay more attention to your own medical problems, and not rely on me to bail you out. If all of my patients did this, there would be chaos!"

It is easy for a telephone call from a patient to be seen as an intrusion into the physician's day (or night). They are usually unscheduled, they may interrupt an already busy day of scheduled office visits, or they come after hours when the physician is taking care of his or her personal life. Add to this the low (if any) rate of reimbursement and the stress of managing a patient without a physical exam, and it isn't surprising that a physician will sometimes have a negative attitude before even picking up the phone to answer a call from a patient. Unfortunately, an angry physician is less able to be objective in taking a history and making management decisions (2,4,15,16).

A number of approaches may be effective in addressing the problem of handling phone calls during the day. These are addressed in greater detail in Chapter 19, but the basic idea is to create a workable system for handling calls, be it triage by a mid-level care provider (e.g., nurse, physician assistant) or designated hours for non-urgent clinical calls. No system, however, will (or should) filter out the most challenging calls: the calls that do not fit neatly into a protocol and need a physician's judgement right away. Crushing chest pain in a 50-year-old may not even require the physician's consultation before calling an ambulance: a nurse performing a triage role can assess this adequately. It is the vague chest pain in a patient with chronic anxiety and somatization that has the physician pulling his or her hair out. To manage these callers properly under optimal circumstances is stressful. To manage them during a busy day taking care of patients, or a busy evening taking care of children, can be nerve-racking.

A physician needs to remember that patients on the other end of the telephone may be just as sick as those in the office or the emergency room, and that they require the same amount of attention and thoroughness. Although the goals of care on the telephone may be different, with a greater emphasis on triage and symptom management than on making a specific diagnosis, the quality of care must be the same. To accomplish this, the physician must do whatever it takes to achieve a mindset that permits his or her best work to be done. This may mean changing rooms, pulling the car to the side of the road, or quickly ruling out medical emergencies and calling the patient back later.

In Case 2-7, Dr. Williams was in a bad mood already when she was further annoyed by the call from Mr. Wallach requesting she do extra work because of his mistake. Her bad mood affected her judgment, and she lost perspective on what is essentially a minor inconvenience for her (calling in a prescription) resulting from a common accident (patient losing an important piece of paper).

Physicians should be aware of when their emotional state is suboptimal for patient care. It may be especially easy to take short cuts or to be less than respectful when talking to someone who is not in the office and who is perceived as an intrusion. Before doing an inadequate assessment or saying something that be might regretted later, however, the physician should identify the barrier and address it.

Dr. Williams should have realized that her bad mood might interfere with her care of the caller, Mr. Wallach. She could have called him back later or simply bit her tongue and ventilated about the annoyance with a colleague afterwards.

The most efficient approach in this case would have been to bypass the phone conversation altogether; Mr. Wallach could have left a message with a nurse that he misplaced his prescription. The nurse would have given Dr. Williams the message, and she could have made the call to the pharmacy at a later time when she wasn't feeling so hectic. Regardless of the content of a call, a disorganized or inefficient approach in responding to calls may produce frustration, distraction, and limited effectiveness. This may occur when physicians take calls that require no medical judgment or when no medical record is available. It may also happen when physicians handle calls at various free breaks between patients, regardless of time available, and do not document the call.

Barriers Arising from the Communication Process

Even when physicians and patients begin their interactions constructively, conflict may arise during the encounter. Patients may perceive the physician as disinterested or they may disagree with the physician's management plan. On the other hand, physicians may become frustrated if the patient's history is communicated in vague or confusing terms or if the patient does not accept the physician's management plan.

SHAME AND HUMILIATION

As Lazare et al have noted, shame and humiliation are overlooked and important emotions experienced by patients (4). Again, without the near physical presence and visual contact of the office visit, it is easier on the telephone for the clinician to be, or seem, aggressive, negative, and demeaning to patients. This may lead to avoidance of medical care and poor adherence to treatment. In Case 2-7, it would not be surprising for Mr. Wallach to think twice before calling again. He might even be so embarrassed at being identified as a "problem" patient that he stops going to Dr. Williams.

As stated above, the best approach is to avoid comments that are likely to be humiliating to the patient. Unfortunately, Dr. Williams did not bite her tongue, and responded to Mr. Wallach with an insulting and possibly humiliating comment. After a negative comment has been made, however, the physician can still undo some of the damage. A straightforward apology and statement of respect may allow the patient to see the slip-up as just that, and not as a statement of the physician's true attitude.

Resolution of Case 2-7

Dr. Williams: "I'm sorry, Mr. Wallach. I've had an unusually hard morning, but I shouldn't take it out on you. I know you take your medical care very seriously and this kind of mistake is unusual for you."

Mr. Wallach: "That's OK, Dr. Williams. I know how busy your office can get."

Sometimes, a physician will recognize that he or she is conveying a negative attitude and can correct it, as Dr. Williams did. Other times, without realizing the reason why, the physician may just notice that the rapport doesn't seem to be developing well or that the patient seems annoyed, or the physician starts to feel negative. As stated above, this can be a sign that the patient has an unstated concern, that the physician is not yet aware of the ARC. The physician should try some of the techniques stated above for uncovering the ARC.

Function 3: Negotiating a Treatment Plan and Providing Education

CASE 2-8 "I Have a Splitting Headache. Should I Get a CAT Scan?"

Mr. Edwards, a computer engineer, calls mid-morning about his wife, who has had a splitting headache since getting up in the morning. Dr. Patelli knows that she gets occasional migraine attacks. In an attempt to give a fellow "scientist" a comprehensive explanation of his wife's illness, Dr. Patelli carefully delineates the differential diagnoses and their symptoms, which include migraine, cluster headache, sinusitis, a cerebral aneurysm, meningitis and brain tumor.

Mr. Edwards doesn't "hear" Dr. Patelli's reassuring statements that some of these problems are very rare and don't even fit his wife's clinical presentation. Later, after the telephone call is over, despite the clinician's clear instructions on management, Mr. Edwards decides to take his wife to the local emergency department to get her a CAT scan of the head so that all these terrifying possibilities can be ruled out.

The third function, sometimes considered the "closing," consists of negotiating a treatment plan and patient education. As Lazare et al indicate in the three-function model, the physician should explain to the caller the diagnostic significance of the problem or symptom in terms of causation, seriousness, treatment, and prognosis (4). As stated above, much of the benefit of the telephone consultation comes from assessment of whether emergency in-person evaluation is needed, and if not, reassurance and advice on symptom management. For the patient to benefit from this, he or she must learn what the possible diagnoses may be, why the physician thinks it is or is not an emergency, and what benefit would be derived from following the physician's recommendation. In person, the patient can be motivated to adhere to treatment by seeing the look of concern on a physician's face or by seeing a diagram illustrating the condition. Over the phone, explanation can be more challenging, and negotiating a plan requires more attention to the patient's verbal and non-verbal responses.

Contracting and Adherence to the Treatment Plan

Because most problems presenting on the telephone are usually immediate, specific, and self-limited, getting agreement and follow through with the patient on management decisions is often easier to achieve than in an office encounter. This, of course, depends on how clearly and simply the physician gives the management plan, how complex the problem and management may be, and what other comorbid illnesses (including various medication regimens) may be present. The physician must know if the caller has understood all of the instructions, is in agreement with the plan, and is able to carry it out.

During the course of the telephone consultation, even while eliciting the history, the physician will be negotiating with the caller until agreement is reached. To avoid misunderstandings and misinterpretations, the physician will also need to "check things out" with the caller during decision making and management. The caller must agree that the course of management is reasonable, appropriate, and that it takes into account the patient's priorities (17,18).

PATIENT EDUCATION

Patient education may appear to be a time-consuming chore in a brief call, but the extra one or two minutes that are involved may lead to patients using the office and emergency room more appropriately and efficiently.

**Table 2-12   General Principles for Providing Patient Education
Over the Telephone**

- Use clear and simple terms; avoid medical jargon

- Avoid over-explanations

- Provide information in small "packets"

- Check patient's understanding frequently (*ask–tell–ask*)

- Ask patients to repeat their understanding of what they've been told

- Ask patients to write down any specific instructions (especially medication
  information)

Contrary to much of the research showing poor recall of key points of diagnosis and treatment in person, patients making calls can generally remember the content of calls that they have initiated (4,11). This is
probably because calls tend to involve urgent requests for help rather than
the more routine office visit for health maintenance or monitoring chronic
disease. Callers also will more consistently take the medication or adhere
to instructions given by the physician. Table 2-12 lists some general principles for providing patient education over the telephone.

USE CLEAR AND SIMPLE TERMS AND AVOID MEDICAL JARGON

Just as in the office, physicians on the phone may fall into the trap of using
medical terminology (e.g., hematocrit) and acronyms (e.g., U/A, UTI,
NSAID). The patient may be too shy, ashamed, or overawed to ask what
these terms mean. Even common words such as "trauma" or "hypertension" are often misinterpreted. Medical jargon includes such simple things
as describing drugs by their action (e.g., diuretics, antihistamine) or describing symptoms or signs (e.g., edema, occult blood, incontinence).

AVOID OVER-EXPLANATION

The physician should be careful to avoid over-explanation and limit the
discussion to the working diagnosis and only those conditions about
which the patient is worried. Compulsive and highly methodical physicians,
such as Dr. Patelli in Case 2-8, may over-explain their thinking. This stylis-
tic characteristic may lead to unwarranted fears when describing possible
diagnostic or therapeutic options to a patient.

PROVIDE INFORMATION IN SMALL PACKETS

The physician should give information in small packets, one idea at a time, checking the patient's understanding before moving on. For example, "Well, there is a small epidemic of these kinds of symptoms, and I think the likeliest cause of your problem is a viral infection. Does that make sense?" Then, "This type of illness gets better by itself, and very rarely turns into anything dangerous. Do you understand what I am saying?"

CHECK THE PATIENT'S UNDERSTANDING FREQUENTLY ("ASK-TELL-ASK")

The physician should ask the patient frequently if he or she understands what the physician is saying, and if it makes sense, with a question such as "Does this make sense to you?"

ASK THE PATIENT TO REPEAT HIS OR HER UNDERSTANDING OF WHAT YOU HAVE SAID

Patients may voice understanding of the physician's words (it's easier to respond "yes" to the question "Do you understand?" than to risk embarrassment and say "no"), but whenever the information is important (and if it's not important, why give it?), the physician should ask the patient, "Can you repeat back to me what I've told you? I want to make sure I've said it clearly." This can be done any time during the interview, but should be considered especially at the end.

ASK THE PATIENT TO WRITE DOWN ANY SPECIFIC INSTRUCTIONS (ESPECIALLY MEDICATION INFORMATION)

The patient should be encouraged to write information down before spending time giving detailed explanations. This is especially important when giving numerical instructions (dosing information, phone numbers, addresses, dates, and times). The act of writing the information down activates the patient's participation and may prompt them to ask about any missing information.

Tasks and Skills Necessary in Treatment Plan Negotiation and Patient Education

A number of elements of content are necessary for effectively closing the telephone encounter. These elements, listed in Table 2-13, are discussed below.

Table 2-13 Important Elements in Effective Treatment Plan Negotiation and Patient Education in Telephone Interviews

---

- Check patient's understanding of the condition
- Explain the nature and seriousness of the problem clearly
- Give a specific opinion about the probable course of the problem
- Assess and address the patient's emotional response to the information being given
- Provide advice on management options, including medication
- Negotiate the plan and establish feasibility
- Offer advice on what to watch for/when to call back
- Make a definite statement about follow-up
- Have the patient repeat instructions
- Offer the opportunity for further questions/Offer to speak with a family member to answer questions

---

CHECK THE PATIENT'S UNDERSTANDING OF AND CONCERNS ABOUT THE CONDITION

By now, hopefully the physician will be aware of any specific worries or concerns that prompted the call. However, as mentioned earlier, these concerns may only surface after the patient has had the chance to talk with the physician a bit. It is useful to address the specific concern first. For example, a worry about not being able to go away on a planned vacation may be more important than what the actual diagnosis is. Also, a patient may be concerned about a specific condition, even if the physician thinks this condition is exceedingly unlikely. A patient worried about a myocardial infarction will not want to hear about antacids before being adequately reassured that the chest pain is not cardiac in etiology.

EXPLAIN THE NATURE AND SERIOUSNESS OF THE PROBLEM CLEARLY

It is very important for the physician to communicate his or her assessment to the caller. Because many calls are made purely for clarification as to whether significant "illness" is taking place, this may be the major contribution the physician makes to the care and satisfaction of the patient. There is a trained and built-in desire in clinicians to solve a clinical problem, but in the telephone consultation there is often no definitive diagnosis.

After explaining the nature of the problem, the physician should provide an assessment of the urgency and severity of the problem. If emergency evaluation is indicated, the patient should be told this clearly. The physician should tell the patient what complications he or she is at risk for and what can be done to avoid them. If appropriate, the patient should be told to call "911," or the physician can do this for the patient.

If the condition does not require emergency evaluation, the physician should explain this and why it is so. Even if the diagnosis is unclear and "watchful waiting" seems the most reasonable strategy, this must be communicated in such a way that the patient understands and feels comfortable with the plan.

GIVE A SPECIFIC OPINION ABOUT THE PROBABLE COURSE OF THE PROBLEM

If the problem is not an emergency, advice on when the symptoms should start to resolve can be quite helpful. The physician can avoid repeat calls by informing the caller of the expected duration and course of the illness.

ASSESS AND ADDRESS THE PATIENT'S EMOTIONAL RESPONSE TO THE INFORMATION BEING GIVEN

In performing Functions 1 and 2, the physician can ascertain the patient's emotional state, which on the telephone is often marked by fear and worry. A physician's attempt at reassurance, however, is not always successful. As in Case 2-8, the physician's explanation of the working diagnosis and plan may actually result in the patient becoming more worried. The physician may not be able to "read" the listener's alarmed reactions to what he/she is saying. Asking directly "How are you feeling about what I'm telling you?" or "Are you still worried?" allows the physician to reassure the patient more effectively.

PROVIDE ADVICE ON MANAGEMENT OPTIONS, INCLUDING MEDICATION

Explain what medications may be helpful and what their desired effect and side effects are. Medications prescribed over the telephone should be the most innocuous possible. Consider using brand names to help the patient find the product on the shelf, but tell the patient when generics would be appropriate. This is a good opportunity to ask about medication allergies, if not already done.

Non-pharmacological treatments should be mentioned specifically and in detail. Recommendations on specific foods or fluids should mention what type of food or fluid is to be avoided or consumed, how often, and for how long. Rest is frequently recommended, but rest for an acute asthmatic is different from that of an ankle sprain.

NEGOTIATE THE PLAN WITH THE PATIENT AND ESTABLISH FEASIBILITY

Just as in the office setting, patients may have strong feelings about their management. On the phone, negotiation can be frustrating for many reasons, such as time pressures and lack of pre-existing doctor-patient relationship. The clinician must be willing to consider other alternatives if the patient is not in agreement. Even patients who do not voice their disagreement may not agree, and may have no intention of following through with the plan, such as Mr. Edwards in Case 2-8. Patients should be asked, "How does this plan sound to you?" or "Does this sound reasonable?". This will give patients like Mr. Edwards a clear opportunity to voice any concerns with the plan. In other situations, the plan may be unacceptable because it is not feasible. Before deciding on the management plan, it is necessary to know if adherence to such a plan would be possible. Without prompting, patients may be reluctant to admit that it is not feasible or practical for them. Can they get a prescription delivered from the pharmacy or get transportation to the emergency department? Again, there is no substitute for directly asking the patient.

OFFER ADVICE ON WHAT TO WATCH FOR AND WHEN TO CALL BACK

The physician should describe which signs would indicate worsening of the condition and how to reach help if that becomes necessary. In addition, the physician should appreciate that in a brief telephone call, the patient may forget to ask certain questions. Whenever appropriate, "leave the door open" for patients to make contact if there are any other questions, if there is a change in their condition, or even if they are "just worried." Mr. Edwards might have taken Dr. Patelli up on such an offer before taking his wife to the emergency department.

MAKE A DEFINITE STATEMENT ABOUT FOLLOW-UP

Tell the patient when he or she should be seen for follow-up and whom he or she should call to arrange this.

HAVE THE PATIENT REPEAT YOUR INSTRUCTIONS

The patient should be asked to repeat back the management plan. This may reinforce particular instructions, such as medication doses or how the illness should be monitored. This also serves to reassure the clinician that he or she has been understood. Again, encourage the patient to write down any specific items of information that are easily forgotten or very important, such as how to use medication or indications for calling back right away. If this had been done in Case 2-8, Mr. Edwards, while repeating back the instructions, might have realized that they were unacceptable to him, and pursued an alternative recommendation from the physician.

OFFER THE OPPORTUNITY FOR FURTHER QUESTIONS

At the end of the call, the physician should provide the opportunity for further questions by saying, "Do you have any questions?" or "Is there anything else you need to talk about?" or "Have I dealt with everything now?"

REFERENCES

1. **Halberstam MJ.** Medicine by telephone: is it brave, foolhardy, or just inescapable? Mo Med. 1977;May 15:11-5.
2. **Curtis P, Evens S.** The telephone interview. In: Lipkin M Jr, Putnam SM, Lazare A, eds.The Medical Interview. Clinical Care, Education and Research. London: Springer; 1995:187-95.
3. **Stewart MA, Buck CW, McWhinney IR.** How illness presents: a study of patient behavior. J Fam Pract. 1978;2:411-4.
4. **Lazare A, Putnam SM, Lipkin M Jr.** Three functions of the medical interview. In: Lipkin M Jr., Putnam SM, Lazare A, eds.The Medical Interview. Clinical Care, Education and Research. London: Springer;1995:3-17.
5. **Perrin EC, Goodman HE.** Telephone management of acute pediatric illnesses. N Engl J Med. 1978;298:130-5.
6. **Katz HP, Pozen J, Mushlin AI.** Quality assessment of a telephone care system utilizing non-physician personnel. Am J Public Health. 1978;68:31-7.
7. **Levy JC, Strasser PH, Lamb GA, et al.** Survey of telephone encounters in three pediatric practice sites. Public Health Rep. 1980;95:324-8.
8. **Durr W.** A call center primer. Healthcare Information Management. 1998;12:5-17.
9. **Morgan WI, Engel GL.** The Clinical Approach to the Patient. Philadelphia: WB Saunders; 1969.
10. **Quill TE.** Barriers to effective communication. In: Lipkin M Jr, Putnam SM, Lazare A, eds.The Medical Interview. Clinical Care, Education and Research. London: Springer; 1995:110-21.
11. **John E, Curtis P.** Physician attitudes to after-hours callers: a five-year study in a university-based family practice center. Fam Pract. 1988;5:168-73.
12. **Cassell EJ.** Untwisting the fibers of paralanguage. Patient Care. 1980;14:186-206.

13. **Cohen-Cole SA, Bird J.** Function 2: building rapport and responding to patients' emotions (relationship skills). In: Cohen-Cole SA, ed. The Medical Interview: The Three Function Approach. St. Louis: Mosby–Year Book; 1991:21-7.

14. **Cassell EJ.** Exploring thoughts that underlie speech. Patient Care. 1980;14:148-60.

15. **Elnicki DM, Cykert S, Keyserling TC, et al.** Issues affecting residents' attitudes about telephone medicine for patients. Teaching and Learning in Medicine. 1996;8:142-7.

16. **Evens S, Curtis P, Talbot A, et al.** Characteristics and perceptions of after-hours callers. Fam Pract. 1985;2:10-6.

17. **Heaton PB.** Negotiation as an integral part of the physician's clinical reasoning. J Fam Pract. 1981;13:845-8.

18. **Davis MS.** Physiologic, psychological and demographic factors in patient compliance with doctor's orders. Med Care. 1986;6:115-22.

3

MEDICOLEGAL ISSUES IN TELEPHONE MEDICINE

Lisa Reisman, Esq.

KEY POINTS

- The trigger to medical malpractice suits is generically known as "a failure to communicate," a problem that is further complicated when medical decisions are made over the telephone and without the visual cues on which physicians normally depend.
- Physicians may delegate non-administrative patient calls to nurses, nurse practitioners, and physician assistants, but never to licensed practical nurses, unlicensed personnel, answering services, or office receptionists.
- With the exception of current patients being seen on a regular basis, physicians should refrain from giving advice over the phone about a course of treatment and should require the patient to come to the office; if the physician opts to make a diagnosis, he or she should inform the caller that it is only presumptive and that a confirmatory physical examination and other testing may be necessary.
- It is legally risky for a physician to fail to return a patient's call in a timely manner. When a call is disconnected or the patient is unreachable, the physician should make a reasonable effort under the circumstances to reach the patient.

Continued

- Because patients' attitudes toward physicians and medical staff are often shaped by the way their telephone calls are handled, anyone on a medical staff who has telephone contact with patients should be adequately trained.
- A written telephone policy or protocol, as opposed to verbal instructions, clarifies for the medical staff what is expected when handling incoming calls and helps to prevent mistakes in judgment that may lead to medical and legal complications. Evidence that the staff followed the applicable protocol also provides solid grounds for a legal defense.
- Complete documentation of a phone call is probably the most effective prophylaxis against potential litigation.
- In the event of an emergency, the practitioner must give either appropriate advice or a referral; termination of the call may be considered legal abandonment.
- Disclosing a patient's diagnosis, treatment, or prognosis to unauthorized parties or on an answering machine constitutes an invasion of privacy and may carry strict statutory penalties, particularly when issues like AIDS, abortion, or drug abuse are involved.
- Because a physician's diagnosis over the phone has not been confirmed by a clinical examination of the patient, medications prescribed over the telephone should be oriented toward symptom relief and be the most innocuous possible.
- Physicians who practice telephone medicine across state lines should apprise themselves of applicable state licensure requirements in order to avoid liability for practicing medicine without a license.
- If the mental capacity of the caller is in question, the physician should attempt to elicit sufficient information to make a recommendation; if such information is not forthcoming from the caller, the physician should do what he or she deems reasonable under the circumstances.

In light of patients' propensity to sue physicians when they are angry or dissatisfied about a treatment outcome, liability should be a matter of concern to all practitioners who give medical advice over the phone, including physicians, nurses, nurse practitioners, and physician assistants.

When used carelessly and without due appreciation for the substantial legal risks it may impose, a phone call with a patient can become a land mine littered with miscommunications, misdiagnoses, delayed treatments, or other unnecessary complications.

Medical practitioners should not be lulled into a false sense of security by the few lawsuits on record against medical professionals in the area of telephone medicine (1). More than ever before, the current health care environment has made the telephone indispensable for patients without the financial resources or insurance to seek clinical evaluations. Even for patients with insurance, the method of payment for regular or emergency care may discourage too many visits. As long as practitioners make judgments and less-than-optimum outcomes occur, litigation will follow. The sharp rise in patients' use of the telephone for medical advice, diagnoses, and treatment will doubtless trigger an increase in lawsuits against medical professionals practicing telephone medicine (2).

The aim of this chapter is to sensitize medical professionals to situations that may lead to legal repercussions when patients seek medical advice over the telephone. It then offers suggestions for feasible precautions that, if followed, will make the practice of telephone medicine more consistent and workable while decreasing the medical professional's legal vulnerability. The scenarios cited at the beginning of each section are designed to raise awareness of worst-case scenarios likely to lead to litigation when a physician fails to adhere to the basic legal safeguards covered in this chapter.

WHAT IS MEDICAL MALPRACTICE?

Medical malpractice, literally, "negligence committed by a professional in the performance of professional duties," is the most common reason for a lawsuit in medical practice litigations (1). Contrary to some popular notions, a physician is not liable for malpractice when there is a bad result or when a patient suffers a complication, either of a diagnostic or therapeutic procedure. Malpractice occurs only in the event that the physician or other medical practitioner fails to consider the possibility that such a complication might occur, fails to watch for it diligently or to recognize it promptly, or to treat it in a timely or appropriate fashion (3).

In other words, medical malpractice occurs when a physician fails to meet a certain standard of care and causes damage to the patient as a result. For any medical professional, including physicians, the standard of care is the care that a medical professional of the same level of expertise and training would deem reasonable and standard under similar circumstances (1). For a nurse dealing with a frantic or an incoherent caller, for example, the standard of care is measured by her ability to assess the caller's condition as accurately as possible, based on her nursing experience and on diagnostic information available in specific protocols, compared with how other nurses with the same level of expertise and training would assess the caller's condition under similar circumstances.

Patients initiate malpractice suits for a variety of reasons, including loss of physician-patient rapport, unrealistic expectations, and the litigious environment in which medicine is currently practiced (4). Although lawyers frequently cite a failure of the medical practitioner to meet the applicable standard of care, the primary catalyst is an outcome that is less satisfactory than the patient expected (5). The trigger to malpractice suits is generically known as "a failure to communicate," a problem only complicated when medical decisions are made over the telephone and without the visual cues on which physicians normally depend (5). This is particularly true when the medical professional making the decision is not sensitive to the potentially serious implications when telephone procedures are not well defined.

WHO SHOULD GIVE ADVICE OVER THE TELEPHONE AND WHO SHOULD NOT?

CASE 3-1 Receptionist Dispenses Advice to a Patient Calling with Headache

A patient calls her physician's office complaining of a crushing headache. The receptionist tells the patient that all medical staff have left for the day but comments, "whenever I have a bad headache, ibuprofen works like a charm." The patient presents to the emergency department the next day in a coma caused by a subarachnoid hemorrhage.

CASE 3-2 Nurse Fails to Question Patient About Relevant History

---

A nurse takes an abbreviated history over the phone, diagnoses a patient with simple heartburn, and tells the patient that calcium carbonate will resolve the discomfort. The nurse, who fails to use a symptom-based protocol, neglects to question the patient about cardiac risk factors. The patient later presents to the emergency department with an acute myocardial infarction.

Physicians commonly delegate patient calls to nurses, nurse practitioners, and physician assistants and allow them to decide what immediate action, if any, should be taken in response to a patient's call. To minimize the risk of liability, the state Nursing Practice Act, which defines the standards of conduct for all persons holding a nursing license, should be reviewed, and organizational procedures should conform to it accordingly (6). In particular, it is essential that the caller understand that the nurse is making an assessment, an evaluation; he or she is not making a diagnosis or providing the result of an evaluation (6). For example, the nurse should preface his or her recommendations by saying, "Based on my assessment of your condition....(6)"

Nurses should make assessments over the telephone only with symptom-based protocols (7) (see Chapter 19). In addition to promoting consistent treatment among patients, symptom-based protocols help protect nurses from charges of practicing medicine without a medical license. Moreover, if litigation arises years later, documented use of protocols can be used to reconstruct what advice was given and to prove that the applicable standard of care was met. Without such documentation, it becomes the word of the nurse against the word of the usually severely damaged, and therefore infinitely more sympathetic, patient in the courtroom.

If the nurse is working with a protocol that does not address a particular symptom described by the patient or deems it necessary to deviate from the protocol in any way, she should contact the physician before proceeding (7). If the physician is not available, the nurse should direct the patient to go to the emergency room or the physician's office.

Giving nursing advice over the telephone requires independent judgment and skill and therefore should not be delegated to licensed practical nurses, unlicensed personnel, answering services, or an office receptionist. These staff members should collect only basic information such as the patient's name, telephone number, and chief complaint, and then put the call through to a triage nurse or physician.

WHEN SHOULD A PHYSICIAN GIVE ADVICE OVER THE TELEPHONE?

CASE 3-3 Physician Fails to Make a Follow-Up Plan

A physician fields a call from a patient he treated 10 years before. The patient complains vaguely of a pain in her shoulder that has lasted for more than 2 weeks. The physician suggests acetaminophen and massage therapy, but, because his schedule is overbooked for the next few weeks, he neither tells the patient to contact his office if the pain continues nor arranges for a follow-up examination. The patient is later diagnosed with breast cancer.

Advice given over the phone is based on the information provided by an untrained observer (the caller). The absence of visual cues presents a challenge to the practitioner's communication skills, with total reliance on verbal communication (see Chapter 2). The caller may not know enough to share the most pertinent and critical symptoms, or may not recognize what is normal and abnormal (1). In short, any telephone assessment is only as good as the information upon which it is based.

Viewed in this light, the most prudent legal recommendation is to avoid giving telephone advice and to inform the caller that a diagnosis cannot be made over the phone. Unfortunately, this is not the most realistic scenario. Telephone advice is an expectation in most practice settings, in cases of real emergencies and when a patient is suffering but is physically unable, or lacks adequate health insurance or financial resources, to undergo a clinical evaluation. In most managed care and private practice settings, telephone advice is now an expected part of care.

The medical professional should be careful not to practice outside of his or her scope of practice or areas of expertise. For the nurse practitioner in private practice, for example, it is advisable to have physician backup, especially for difficult or nonadherent patients (1).

Once a medical practitioner starts giving advice to the caller, he or she assumes a legal duty to the caller. Specifically, for legal reasons, a relationship with the patient is established by a telephone call when advice is given about a course of treatment and it is foreseeable that the patient will rely on that advice (1).

When deciding whether it is appropriate to give advice, the status of the patient should be considered (1). If the caller is a current patient being seen on a regular basis, advice may be fairly simple. If the medical practitioner chooses to make a diagnosis, he or she should, as a rule, inform the caller that it is only presumptive and that a confirmatory physical examination and laboratory testing may be indicated.

If the caller is a current patient but has not been seen in some time, or is not one of the practice's regular patients, it is advisable to limit telephone advice and require the patient to come to the office. If the caller is new to the practice, the physician should encourage the caller to make an appointment for a baseline examination in the office.

WHEN MUST THE TELEPHONE BE USED?

CASE 3-4 Failure to Alert Patient of Lab Report

The medical staff does not alert a patient after her lab report identifies a serious problem until 2 weeks after the lab report is issued. Meanwhile the patient is admitted to a nearby hospital with an upper gastrointestinal bleed.

The telephone also poses legal pitfalls when it is not used (8). Physicians may be found negligent for not returning a patient's call in a timely manner or when a problem is identified and the length of time before the patient is contacted and evaluated is prolonged. Nurses who fail to report to a physician important changes in a patient's condition also may be charged with negligence. To reduce claims of this nature, well-defined written telephone procedures and clear communication between nurses and doctors are crucial.

Follow-Ups

Every call must have some plan for follow-up. The patient should be advised how soon and where he or she must be seen and with whom an appointment should be made. Alternatively, if the circumstances dictate, the patient should be told when the medical practitioner will call to follow up or when the patient should call the medical practitioner. The follow-up plan should always be documented.

Disconnected Calls

At the beginning of each call, the physician or other practitioner should take down the location of the patient and the telephone number from where the patient is calling (6). If the call is disconnected, the practitioner should attempt to recontact the patient. If the practitioner is unable to do so, he or she should assess the patient's condition based on whatever information has been received, both during the call and, if applicable, from previous visits. The more serious the condition potentially may be, the more diligence the physician must practice to reconnect with the patient.

The Unreachable Patient

If the patient cannot be reached, for example, when a physician receives a call from his answering service reporting that a patient has called in describing an adverse medical condition or if lab results show an abnormal condition ("panic values"), the physician should refer to other numbers the patient may have provided. If those efforts fail, the physician should use an overnight mail service or telegram to deliver a letter urging the patient to contact the physician's office, or call the police or 911, as appropriate. In all circumstances, the seriousness of the patient's condition should dictate the aggressiveness of the measures taken by the physician to reach the patient.

WHAT IS IMPORTANT IN TRAINING THE MEDICAL STAFF?

CASE 3-5 Nurse Fails to Answer an Emergency Call

A receptionist who is chatting with a triage nurse about a personal problem lets the phone ring 26 times before picking up. The caller, who has been having severe chest pain, loses consciousness as the nurse picks up the phone.

CASE 3-6 Patient Verbally Abuses Receptionist

A caller is verbally abusive towards a receptionist after being put on hold for 15 minutes. The receptionist haughtily says to him "How dare you talk to me like that!" and hangs up.

Patients' attitudes toward physicians and medical staff are often shaped by the way telephone calls are handled (9). In general, if the physician and staff seem concerned, knowledgeable, and well-organized, the caller is much more likely to be satisfied with the telephone service even if the clinical outcome is not favorable.

Anyone on a medical staff who has telephone contact with patients should be adequately trained (9). Training should cover general interviewing etiquette and techniques, assessment procedures, and the content of telephone advice protocols and should include practice sessions with on-the-spot feedback (6). Once nurses or receptionists are on the job, supervisors should occasionally oversee the calls.

Non physician medical professionals also should be trained in how to handle special situations, such as the suicidal caller, the unconscious patient (the caller may need to be instructed in emergency CPR procedures), the talkative caller, the abusive caller, the caller who is difficult to understand, or calls from the relatives of the patient (6).

The behavior of the receptionist, who usually represents the entry point for the caller, sets the mood for contact with the caller. It is essential that the receptionist have proper motivation, tact, and adequate factual information. Answering the telephone promptly or, if that is not possible, apologizing to the caller earnestly in the event of a delay, is a simple convention that can go a long way in satisfying the patient and decreasing the potential for legal problems in the future.

WHAT WRITTEN MATERIALS SHOULD BE USED BY THE MEDICAL STAFF?

CASE 3-7 Receptionist Inappropriately Directs Call

The office manager verbally instructs a newly-hired receptionist to forward all urgent calls to Nurse X, all routine calls to Nurse Y, and any requests for referrals to Nurse Z. The parent of a diabetic child patient calls, reporting that her child has just fainted. The receptionist panics, forwards the call to the wrong nurse, and the parent is put on hold for 5 minutes. The patient arrives at the emergency department an hour later in a coma.

A written telephone policy or protocol clarifies for the medical staff what is expected when handling incoming calls and helps to prevent mistakes in judgment that may lead to medical and legal complications. Moreover, evidence that the staff followed the applicable protocol will provide solid ground for a legal defense (8). Verbal instructions are not adequate because the staff may ignore, forget, or alter the instructions.

Copies of a printed standard operating procedure, updated as needed, should be given to everyone who uses the office telephones. There should be a master copy, signed by the appropriate physician, on file (7). In addition, the protocols should be reviewed regularly by an attorney to ensure that they are consistent with current standards of care (7).

A receptionist's manual is helpful. It should contain, among other information, regular office hours, transportation instructions to reach the office, the location and telephone number of the preferred emergency department, a listing of telephone numbers of physicians and medical staff and locations where they can be reached in the event of emergency, and triage instructions for incoming calls, including calls requiring immediate attention, calls requiring attention in the near future, and routine calls.

WHICH INFORMATION MUST BE DOCUMENTED?

CASE 3-8 Physician Fails to Document a Call from a Suicidal Patient

While at a dinner party, a physician returns a call to a severely depressed patient. After a lengthy discussion, the physician determines that the patient is not at risk for suicide and provides a recommendation for follow-up the next day. The physician returns to dinner and forgets to document the call. A week later he learns that the patient is in the emergency department after taking an overdose. The patient's family initiates a lawsuit against the physician.

Complete documentation is probably the most effective prophylaxis against potential litigation. Although writing up a call may seem burdensome, it need not be if only the crucial elements are included (Tables 3-1 and 3-2). Even notes made on an index card or on a sheet backed by adhesive and pressed into place on a chart can help indicate that the physician or medical staff acted properly.

Lawyers tend to be deterred by good records. If a caller sues claiming he was given improper advice, a complete record of the conversation can help to minimize the risk of liability. Conversely, a lack of documentation makes the practitioner vulnerable in the courtroom because he is forced to reconstruct the advice he allegedly rendered without concrete proof and because the failure to document the conversation creates the general appearance of carelessness and neglect.

Table 3-1   Elements of a Well-Documented Telephone Record

- Date and time of call
- Patient's name with spelling confirmed
- Date of birth
- Phone number and location
- Chief complaint
- Relevant history, including allergies, recent injuries, past medical history, and current medications, with both positive and negative responses noted
- Assessment
- Follow-up plan
- Signature

Table 3-2   Standard Rules for Documentation

- Write legibly, in ink
- Quote the patient and advice given whenever possible
- Correct mistakes with a single strike-through and label as "error"
- Date and initial the corrected mistake
- Use concise, concrete terminology
- Record information on preprinted forms

At the very least, the practitioner should document the date and time of the call, the nature of the problem (quoting the caller whenever possible), and the advice given. Any documentation, however sparse, should be signed or initialed by the practitioner. If litigation arises years later, a document without a signature identifying the practitioner will likely be useless, particularly if the practitioner no longer is an employee of the institution or office through which he or she treated the patient who is now filing a lawsuit.

If the situation allows, the practitioner should note the caller's name and phone number. If that is not possible, for example, if the call is disconnected before the practitioner has an opportunity to ask for that information, there is nothing more to do except to document the conversation, using quotes whenever possible.

Any time a non-physician practitioner deviates from a protocol, he or she should record the reasons for this action, the discussion with the supervising physician, and any actions taken under the physician's direction (7).

Telephone consultations that physicians have with other physicians should also be noted in the patient's chart, particularly when the consultation influenced patient management.

Documentation logs should be saved for 7 to 10 years, depending on the applicable state statute of limitations (1).

THE IMPORTANCE OF DOCUMENTATION

In one trial, a jury found an office nurse, among others, negligent because of her failure to follow established guidelines in the documentation of her telephone conversation with the parent of an infant patient (5). The nurse had used vague terms in her description of the infant's condition, such as "weak," "listless," and "high fever." In addition, the nurse failed to enter in the telephone log her offer to make an appointment for the infant that day, an offer that the parent declined; the follow-up instructions simply stated "call in AM." The infant was ultimately diagnosed with meningitis and suffered profound neurological sequelae.

Had the nurse simply noted her offer to make an appointment that day and had she described the patient's condition with more specificity, it is likely that she could have avoided or at least minimized her liability. Instead, the jury found the nurse, among others, liable to the patient for $2.5 million.

WHAT SHOULD PRACTITIONERS DO WHEN FIELDING CALLS THAT PRESENT REAL OR POTENTIAL EMERGENCIES?

CASE 3-9 Possible Failure of a Nurse to Adequately Manage an Emergency Call

A patient's husband calls his wife's physician after she has suffered a seizure and is lying on the kitchen floor unconscious. The nurse to whom the call is referred advises the husband to call 911 and hangs up.

Failure to give adequate warning is a prevalent theory in lawsuits involving telephone medicine (1). Although the caller is accountable for following the advice, it is the responsibility of the medical practitioner to convey the urgency of the situation in clear and unambiguous terms and to ensure that the caller understands the consequences of noncompliance (7).

In the event of an emergency, the practitioner must give either appropriate advice or a referral because termination of the call may be considered abandonment (1).

When dealing with a real medical emergency, it is advisable for the medical practitioner to obtain the caller's address and keep the caller on the line while either the medical practitioner or a colleague calls 911 and for the medical practitioner to stay on the line until emergency technicians get to the caller (7).

To convey the urgency of the situation, the practitioner may need to give a working diagnosis. In those circumstances, the words used should be chosen carefully to avoid giving an actual medical diagnosis. The practitioner might say, for example, "your symptoms suggest that you may have an injury to your spine" (7).

If the condition is less urgent but emergency treatment is still indicated, the practitioner should state instructions clearly and give a specific time frame in which to seek medical attention in person. It is not enough to recommend that the caller go to the emergency department (7). The practitioner should recommend that the caller go the emergency department within the next hour or serious injury or death could result (7).

HOW CAN PRACTITIONERS AVOID BREACHING THE PATIENT'S CONFIDENTIALITY AND RIGHT TO PRIVACY?

CASE 3-10 Staff Member Gives Confidential Information to a Non-Family Member

---

A staff member at a hospital's information desk receives a call from a man inquiring about the condition of a patient who suffered a gunshot wound. The staff member explains that hospital policy does not allow releasing such information and asks the caller if he is a relation of the patient. The caller replies, "No, but he's a friend of my daughter." The staff member discloses the patient's condition.

Disclosing a patient's diagnosis, treatment, or prognosis to unauthorized parties constitutes an invasion of privacy. When issues like AIDS, abortion, or drug abuse are involved, safeguarding confidentiality is particularly important (8). States have enacted stringent laws to govern such sensitive circumstances. With respect to HIV, state laws regarding disclosure and the duty to warn potential contacts vary and depend on the state in which one practices. Before implementing a policy for patients with HIV infections, an attorney should be consulted.

To avoid a potential breach of confidentiality, the physician or medical staff should, whenever possible, talk directly with the patient. If the patient is not immediately available, the physician or medical staff should leave a message for the patient to return the call. Medical information should never be left on an answering machine or with someone other than the patient, even the patient's spouse, except with the patient's specific written permission.

It is safe to convey patient information to a family member or friend who has been expressly designated by the patient as authorized to receive such information, as well as to health care professionals directly involved in the patient's treatment (8). However, representatives of insurance companies and other agencies with a legitimate interest also may call for updates on patients' status. Under certain circumstances, an employer may have the right to verify the condition of an employee receiving care.

In responding to all requests for medical information other than those by the patient, it is essential to ask for and record the name of the caller and his relationship to the patient before releasing any information. If there is any uncertainty about the identity of the caller or whether he or she has the patient's permission or the legal right to confidential information, it is advisable not to provide the information. In such circumstances, the physician may request that the patient's written consent to disclosure of information be mailed to the office.

WHAT CONSIDERATIONS ARE IMPORTANT WHEN PRESCRIBING MEDICATION?

CASE 3-11 Physician May Have Prescribed Too Much Medication

A patient comes to the clinic complaining of symptoms suggestive of an excess medication dosage. The physician expresses surprise because, according to the patient's chart, he had placed the patient on a normal dosage at the previous visit. The patient reminds the physician that he had told the patient to double up on pills when the patient called the week before. The physician recalls telling the patient to take an extra pill for 3 days only, but he forgot to document this.

A large percentage of telephone work involves refilling prescriptions, particularly those for chronic diseases. Miscommunication and mistakes, such as when calling in prescriptions for a patient to a pharmacy or when noting medication in the patient's chart, can have serious consequences (8).

Because a physician's diagnosis over the phone has not been confirmed by a clinical examination of the patient, medications prescribed over the telephone should be oriented toward symptom relief and be the most innocuous possible. If the caller requesting the prescription is not the physician's patient, the physician should prescribe the smallest quantity possible, to cover a weekend, for example, and advise the caller to follow up with his or her primary care physician. Over-the-counter drugs should be used when they accomplish the same objective as prescription drugs. They are easier to obtain and save the physician phone calls to the pharmacy.

Before suggesting medication, the physician should question the patient about allergies or a history of adverse side effects and which medications the patient is currently taking. After the medication is prescribed, it should be entered into the patient's permanent record.

Because many medications sound the same, the physician as a rule should insist that the pharmacist taking the order repeat back to him the spelling of the medication, the dosage, and the applicable regimen. The physician should also review with the patient the medication regimen and, if there is any question of patient confusion, ask the patient to repeat back the physician's directions. The physician should note in the patient's record that such a communication occurred, for example, "Patient understood dosage and regimen."

Physicians should consider authorizing an experienced physician assistant, nurse practitioner, or nurse to assume responsibility for refilling medications and reviewing their actions afterwards. In many states, this form of delegation is permitted by law and can prove a time-saver. If the physician opts to make such a delegation, a written policy should include a list of medications, including narcotics, which should rarely, if ever, be ordered by phone (9). Another technique is to establish a telephone line on which patients can record their refill requests on voice mail for later action (9).

HOW CAN PHYSICIANS PROTECT THEMSELVES LEGALLY WHEN PRACTICING TELEPHONE MEDICINE ACROSS STATE LINES?

CASE 3-12 Physician Renders Advice to Patient in State in Which He Is Not Licensed

A physician licensed to practice medicine in State A renders advice to a caller in State B without verifying that it is legally permissible for an out-of-state physician to do so. The physician is sued in State B for practicing medicine without a license.

Physicians who practice telephone medicine across state lines should be aware of applicable state licensure requirements in order to avoid liability for practicing medicine without a license. Some states require licensure of any physician providing medical advice to a patient located in that state, even when the physician is providing the telephone services from another state in which he or she is licensed (10).

WHAT SHOULD PRACTITIONERS DO IF THE CALLER'S DECISION-MAKING CAPACITY IS IN QUESTION?

CASE 3-13 Physician Fails to Recognize Emergency Involving a Delusional Patient

---

A patient calls a physician babbling incoherently about monsters eating up his chest. When the physician asks the patient where he is, the patient gives him his address. The physician hears a loud thud and then only silence. The physician hangs up, failing to call 911.

Our legal system endorses the principle that all persons are competent to make reasoned decisions unless demonstrated to be otherwise (11). One of the most complex issues facing physicians is the management of medical treatment when a patient's rational decision-making ability is in question: for example, if an individual is delirious or his or her consciousness is otherwise impaired, has a significant thought disorder, has deficits in short-term memory, or has a mood disorder that impairs decision making.

The term "capacity" is often mistaken for "competence." Physicians determine whether a patient has "capacity" to make a rational medical decision. In the event that an official declaration is needed, "competence" is determined in a legal proceeding once a patient is deemed to lack capacity. Because of the cumbersome and often costly nature of a competency procedure, patients deemed to lack capacity are considered de facto incompetent in a medical setting.

The standards relevant to the assessment of decision-making capacity can vary from state to state. Generally, however, any reasonable assessment of capacity must address the patient's ability to comprehend the prevailing medical condition and the available treatment options, as well as their risks and benefits. Research shows that primary care physicians are increasingly requesting psychiatric consultations to make this determination in order to protect themselves from a medical-legal perspective. Physicians may be exposed to liability for either providing or refusing to provide medical treatment for a patient later deemed incapable of making a reasoned medical decision.

On the telephone, a physician's resorting to a psychiatric consultant when he or she suspects that a caller lacks decision-making capacity is usually not realistic. In such circumstances, however, there are certain protocols a physician may follow to assist the patient medically and reduce potential liability. Initially, the physician should attempt to elicit from the caller the address and the telephone number from which he or she is calling. If the physician can obtain sufficient information from the caller to make a recommendation, he or she should ask the caller to repeat back the recommendation. If the physician cannot elicit sufficient information or if the patient's response demonstrates that he or she does not understand the physician's recommendation, the physician should ask the caller if there is any family member or friend who can provide information about the patient's condition. If the caller indicates that there is no one in the immediate vicinity, the physician should do whatever he or she decides is reasonable under the circumstances.

In some cases, it may not be possible for the physician to determine the location of the patient or to otherwise communicate with him or her in a meaningful way. In such cases, the physician should simply record the call in the medical record, noting responses to inquiries in quotation marks whenever possible.

Under the legal doctrine of "implied consent," the physician can undertake a treatment intervention that is life-sustaining when a de facto incompetent patient's medical condition has imminently dangerous consequences. This doctrine assumes that the physician is acting on behalf of the incompetent patient in a manner consistent with what any reasonable person in that emergency situation would prefer. It is unclear how "implied consent" applies to emergency situations unfolding over the telephone, particularly when the caller's competence has not been determined. However, if the condition of the caller is life-threatening and the physician has the caller's address or phone number, or some idea of where the caller is, he or she should call 911.

REFERENCES

1. **Robinson DL, Anderson MM, Erpenbeck PM.** Telephone advice: new solutions for old problems. Nurse Pract. 1997;22:179-80, 183-6, 189.
2. **Phelan JP.** Ambulatory obstetrical care: strategies to reduce telephone liability. Clin Obstet Gynecol. 1998;41:640-6.
3. **Gregory DR.** Medical malpractice prevention. Leg Med. 1982;177-86.
4. **Huycke LI, Huycke MM.** Characteristics of potential plaintiffs in malpractice litigation. Ann Intern Med. 1994;120:792-8.
5. **Katz HP, Wick W.** Malpractice, meningitis, and the telephone. Pediatr Ann. 1991;20:285-9.
6. **Bartlett EB.** Managing your telephone liability risks. J Healthcare Risk Management. 1995;15:330-6.
7. **Gobis LJ.** Reducing the risks of phone triage. RN. 1997;60:61-3.
8. **Tammelleo AD.** Staying out of trouble on the telephone. RN. 1993;56:63-4.
9. **Johnson BE, Schmitt BD, Wasson JH.** Taming the telephone. Patient Care. 1995;29:136-56.
10. **ACOG Committee on Professional Liability.** ACOG Committee Opinion. 1999; No. 221.
11. **Leo RJ.** Competency and the capacity to make treatment decisions: a primer for primary care physicians. Primary Care Companion. J Clin Psychiatry. 1999;5:131-41.

PART II

PRACTICING TELEPHONE MEDICINE: THE CLINICAL SETTING

4

....

CHEST PAIN

....

David L. Stevens, MD

KEY POINTS

- The primary role of the telephone consultation for patients with chest pain is to assess the likelihood of acute coronary syndrome or another life-threatening condition and facilitate appropriate evaluation and treatment of patients considered to be high risk.
- Rapid intervention is especially important in possible acute myocardial infarction because early treatment (within 1 hour when possible) decreases mortality and long-term morbidity.
- Chest pain that is pleuritic, sharp or stabbing, positional, or reproducible by palpation is less likely to be an acute coronary syndrome.
- Chest pain with radiation to either or both arms, nausea or vomiting, diaphoresis, or chest pain that is the patient's most important symptom is associated with an increased probability of acute coronary syndrome. These symptoms should be asked about before a patient is designated as low risk (not needing immediate emergency department [ED] evaluation).

Continued

- A past history of vascular (cardiac, cerebral, or peripheral) disease is associated with a much higher risk of acute coronary syndrome. These co-morbidities should prompt a lower threshold for referral.
- Physicians should exercise greater caution with older patients because the risk of mortality from acute myocardial infarction increases with age.
- Female gender does not lower the probability of acute coronary syndrome in patients with chest pain. Physicians should be careful to avoid gender bias.

When a patient calls with chest pain, the physician's immediate concern is whether the pain could be an acute coronary syndrome. The risk involved with acute coronary syndrome (either acute myocardial infarction or unstable angina) dictates that the physician must focus on the critical decision of triage. As discussed below, acute coronary syndrome is common, potentially fatal, and often treatable when recognized promptly.

Although this chapter will focus on acute coronary syndrome, chest pain can also be a manifestation of other serious or life-threatening cardiac and pulmonary conditions. Chest pain often has a less dangerous cause such as gastroesophageal reflux, muscle strain, or anxiety/panic disorder, but it is the worry about the more dangerous conditions that keeps the awakened on-call physician from falling back to sleep.

Does this mean every patient with acute chest pain should be immediately taken by ambulance to an ED? Clearly, the answer is no. Most physicians have evaluated patients with chest pain in whom they could determine, by history alone, that a life-threatening condition was very unlikely.

Classic teaching on the diagnosis of chest pain cites the history as the most important component in the evaluation of patients with chest pain (1). Much has been written describing the typical symptoms of the patient with acute coronary ischemia (1). Relatively few studies are available to support the classic teachings, but a small number of studies have been performed that demonstrate the utility of the history in assessing the probability of acute coronary syndrome. These findings (results of studies in EDs) must be applied with caution when evaluating a patient over the telephone. A physician managing a caller with chest pain must combine skilled history taking with knowledge of these studies and the classic teachings to be able to provide the best advice to the caller.

This chapter will focus on answering the following important questions:

1. Does the patient require emergency evaluation?
2. What other emergency conditions should be considered if the patient is determined to be at low risk for an acute coronary syndrome?
3. How should the patient with low-risk chest pain be managed?

Given the particular urgency of the symptom of chest pain, and the fact that non-physician office staff initially receives many patient calls, this chapter will also address the question:

4. How should non-physician staff manage calls for chest pain when a physician is not immediately available?

BACKGROUND

Epidemiology

Coronary artery disease is the most common cause of death in men and women in the United States, causing one in five deaths (1,2). Approximately 30% of people who suffer an acute myocardial infarction will die from it, approximately half of these deaths occurring within the first month (1,3). The risk for death from a myocardial infarction is especially high in elderly patients; the mortality rate from acute myocardial infarction increases with age without any specific threshold (see discussion of role of risk factors below) (4). The main cause for concern in unstable angina is the one in three risk of progression to acute myocardial infarction.

There are no studies assessing the incidence of acute coronary syndrome in patients calling with chest pain. If the statistics from ED studies can be used as a guide, the rate of acute coronary syndrome in patients with acute chest pain is high: 15% have acute myocardial infarction, whereas 30% to 35% have unstable angina (5).

Utility of Early Diagnosis

The primary purpose of early diagnosis of acute coronary syndrome is that rapid treatment can reduce mortality while limiting myocardial damage. Mortality reduction in acute myocardial infarction, for example, can be achieved with administration of beta-blockers, angiotensin-converting enzyme inhibitors, and aspirin (3,6). Furthermore, rapid reperfusion with thrombolytics (or angioplasty) decreases mortality further, perhaps by as much as 50% (1). The effect of thrombolytic treatment is greatest when initiated within 1 hour after the onset of symptoms (7). Unfortunately, nearly two-thirds of patients hospitalized with acute myocardial infarction do not receive thrombolytic therapy, largely because many patients do not seek treatment promptly (8). Clearly, callers who may be having an acute coronary syndrome need to be treated in an ED or equivalent setting as soon as possible.

The importance of proper triage is demonstrated by a study that showed that patients mistakenly discharged from an ED with a missed diagnosis of acute myocardial infarction had a mortality rate of approximately 25%, which is nearly double the mortality rate expected had they been admitted (5). Furthermore, missed diagnoses of acute myocardial infarction result in the most costly category of malpractice litigation in the ED (5).

On the other hand, unnecessary ED referrals have disadvantageous consequences. Ramifications of referring patients with chest pain with a benign etiology to the ED include causing the patient needless worry, delay in initiating symptomatic treatment, possible exposure to risks of treatment, and cost. Every year in the United States there are 5 million ED visits for chest pain, with 1.5 million hospital admissions to rule out acute coronary syndrome. The annual direct cost of this is approximately $3 billion (9). Even a small reduction in the rate of unnecessary evaluations might result in great savings to an overextended medical system.

LIMITATIONS OF APPLYING EMERGENCY DEPARTMENT STUDIES TO TELEPHONE MANAGEMENT

Although many studies demonstrate the utility of the medical history in ruling out acute coronary syndrome (primarily acute myocardial infarction), these studies have limitations (Table 4-1). Most of the clinical trials discussed herein studied acute myocardial infarction, not unstable angina. Although the characteristics of the chest discomfort in these two conditions are generally felt to be quite similar (1), this remains a significant limitation to the application of the study findings and necessitates greater reliance on clinical judgment. Whenever managing a potentially life-threatening problem such as chest pain, a physician must exercise great caution in applying the results of studies. The limitations listed in Table 4-1 provide a good argument for the need to perform a careful evaluation of every caller and not to over-rely on a few symptoms to rule out the condition. Similarly, it is not advisable to use protocols such as those presented in this chapter as a replacement for a thorough evaluation by a skilled, experienced physician. Protocols designed for practitioners such as nurses are generally more conservative, advising ED referral in most cases (see algorithm on p 103).

Table 4-1   Limitations of Studies Assessing Utility of Historical Factors in Diagnosis of Acute Myocardial Infarction

Characteristic of Study	*Limitations*
ED setting (10)	• Risk of AMI in callers may be higher or lower • Study patients are screened (vital signs, general inspection), possibly leading to lower-risk study population
Exclusion criterion: abnormal chest radiograph (10)	• Patients with heart failure excluded from study, leading to lower risk study population • Dangerous pulmonary pathology ruled out, which cannot be done over the phone
Poor inter-observer reliability: different interviewers had high rate of differing responses to same questions regarding characteristics of chest pain (11)	• Reliability of information gathered over the phone may not be acceptable

DOES THE PATIENT
NEED TO BE REFERRED TO
AN EMERGENCY DEPARTMENT
IMMEDIATELY?

A significant challenge in the management of acute coronary syndrome over the phone is that no single finding on history, or even a group of findings, has been demonstrated to rule in or out acute coronary syndrome with sufficient accuracy (10). Physicians must get a full picture of the patient's history and apply clinical judgement. This section presents a logical method of organization for gathering this full picture and summarizes the available evidence for each element of the history.

What Questions Should Be Asked About Symptoms?

Information that will help the physician decide on how to manage callers with chest pain falls into two categories: 1) characteristics of the patient's symptoms and 2) risk factors for acute coronary syndrome/coronary artery disease (Table 4-2).

Table 4-2   Important Questions for Callers with Chest Pain

Character of Symptoms

1. *What exactly are you feeling?* (Quality of the pain)

2. *Where are you feeling this?* (Location of the pain)

3. *When did it start? How long does it last?* (Chronicity of the pain)

4. *What makes it better? Worse?* (Alleviating/exacerbating factors)

5. *What else are you feeling?* (Associated symptoms)

Information Regarding Risk Factors

1. *Have you ever been told you have heart disease?*

2. *Other medical problems?*

3. *Any medications? Illicit drugs?*

4. *How old are you?*

The first task is to gather a full description of the symptoms. Because subtle differences in the characteristics of the discomfort may affect the likelihood of serious disease, the patient should first be asked to describe the symptoms in his or her own words. Important areas to encourage the patient to focus on include the quality, location, and chronicity of the pain; alleviating/exacerbating factors; and associated symptoms (1).

Quality of the Chest Pain: "What Exactly Are You Feeling?"

Chest pain that is pleuritic, sharp or stabbing, positional, or reproducible by palpation has a lower probability of acute coronary syndrome (10). Table 4-3 lists likelihood ratios for each of these symptoms.

One study found that the combination of positional chest pain, reproducibility by palpation, and chest pain that was pleuritic, sharp, or stabbing lowered the risk of acute myocardial infarction enough for an electrocardiogram to be unnecessary (11). An abnormal electrocardiogram in these patients would in all likelihood be a false positive. Unfortunately, restrictive exclusion criteria (patients entered the study only after a normal chest x-ray) prevent applying this rule to patients evaluated over the telephone. Moreover, the presence or absence of reproducibility with palpation may be less reliable when evaluated by the patient than by the physician palpating the painful area.

Table 4-3   Symptoms Associated with a Lower Risk of Myocardial Infarction and Their Likelihood Ratios

Symptom	*Likelihood Ratio (95% Confidence Interval)*
Pleuritic chest pain	0.2 (0.2-0.3)
Chest pain sharp or stabbing	0.3 (0.2-0.5)
Positional chest pain	0.3 (0.2-0.4)
Chest pain reproduced by palpation	0.2-0.4*

* In heterogeneous studies the likelihood ratios are reported as ranges.
Data from Reference 10.

The classic description of angina pectoris consists not of "pain" but of "pressure," "squeezing," "strangling," or even "burning" (1). Clinical trials support this description indirectly; symptoms deviating from this classic description are indeed associated with lower risk of acute coronary syndrome.

Location of the Pain: "Where Are You Feeling It?"

Radiation to either or both arms has been shown to increase likelihood of acute myocardial infarction (Table 4-4) (10), although the radiation may not be present in unstable angina. Substernal location is more typical of coronary ischemia (12).

Chronicity of the Pain: "When Did It Start/How Long Does It Last?"

Anginal discomfort typically lasts 5 to 30 minutes; symptoms for more than 30 minutes suggest infarction (1). Persistent pain lasting a day or longer is less likely to be ischemia or infarction, as is pain lasting only a few seconds.

Table 4-4   Symptoms Associated with an Increased Risk of Acute Myocardial Infarction and Their Likelihood Ratios

Symptom	Likelihood Ratio (95% Confidence Interval)
Radiation to right arm	2.9 (1.4-6.0)
Radiation to left arm	2.3 (1.7-3.1)
Radiation to both left and right arms	7.1 (3.6-14.2)
Chest pain most important symptom	2.0*
Nausea or vomiting	1.9 (1.7-2.3)
Diaphoresis	2.0 (1.9-2.2)

* Data not available to calculate confidence intervals.
Data from Reference 10.

Alleviating/Exacerbating Factors: "What Makes It Better? Worse?"

As mentioned above, pain that can be reproduced by palpation is less likely to be an acute coronary syndrome. Pain that does not worsen with exertion is less likely to be caused by cardiac ischemia (12). Antman and Braunwald have stated that chest pain brought on by excitement/stress, cold weather, or eating a meal is consistent with cardiac ischemia, but, in fact, these factors are clearly associated with non-cardiac chest pain etiologies as well (1).

Associated Symptoms: "What Else Are You Feeling?"

Nausea, vomiting, and diaphoresis are all associated with a higher probability of acute myocardial infarction (Table 4-4) (10), but these may be absent in unstable angina. Chest pain with dyspnea is worrisome, suggesting congestive heart failure or a primary pulmonary process such as pneumonia (see Chapter 7).

What Risk Factors Affect the Likelihood of Acute Coronary Syndrome in Callers with Chest Pain?

The well-known risk factors for coronary artery disease are useful in the general population, but they are actually less useful in determining the risk for an individual patient experiencing chest pain. When these risk factors are applied to the subpopulation of people presenting with acute chest pain, studies do not support their utility in making management decisions. Only history of myocardial infarction has been shown to predict a higher rate of acute myocardial infarction in patients with chest pain, with a likelihood ratio found in various studies to be 1.5 to 3.0 (10). Others have argued that patients with a history of any atherosclerotic disease (coronary peripheral artery or cerebrovascular disease) should also be considered very high risk based on epidemiological evidence; patients with a history of atherosclerotic disease are five to seven times more likely to experience myocardial infarction and subsequent death than the general population (8).

Age is another risk factor that needs to be considered in making management decisions. Older age is associated with an increased risk of complications from acute coronary syndrome rather than an increased risk of acute coronary syndrome itself. Age has not been found to alter the probability that an adult with chest pain is having an acute coronary syndrome. For example, one study found patients with chest pain who are 60 years old or younger have the same risk of acute myocardial infarction as those older than 60 (10). However, there is good reason for being especially careful with older patients because the mortality rate with acute myocardial infarction increases with age (4,13). Although this increase in mortality rate may be partly caused by lower rates of aggressive treatment with thrombolytics, emergency catheterization, percutaneous transluminal coronary angioplasty, and coronary artery bypass grafting (13), these data nonetheless dictate an added degree of caution in the management of older patients with chest pain.

Physicians may choose to ask about the presence of other risk factors such as hypertension, hyperlipidemia, or positive family history. Presence of these risk factors may tip the scales towards immediate evaluation in an otherwise borderline case. However, absence of these risk factors should not be cause for reassurance.

A pitfall with querying about risk factors is that the absence of risk factors does not significantly lower the risk that a patient's chest pain represents an acute coronary syndrome. Coronary artery disease is so common that patients with no risk factors at all remain at significant risk if their presentation is otherwise suggestive of acute coronary syndrome. Therefore, with the exception of personal history of vascular disease (coronary, peripheral, or cerebral) and age, there is no evidence to support using the presence or absence of other risk factors in guiding management of patients with acute chest pain.

With this in mind, the following questions are useful because they may result in lowering the physician's threshold for in-person evaluation.

"Have You Ever Been Told You Have Heart Disease? Have You Ever Had a Heart Attack?"

As stated above, a past history of myocardial infarction is associated with an increased probability that the current episode of chest pain is also a myocardial infarction (likelihood ratio 1.5 to 3.0) (10). Other authors suggest immediate evaluation of patients with any history of coronary artery disease (regardless of whether the patient actually had a myocardial infarction) (4).

Patients with pre-existing angina should be asked about changes in the symptoms, whether the angina is significantly worse than before or comes on at rest or with less exertion than before. These findings suggest an acute coronary syndrome.

"Do You Have Any Other Medical Problems?"

Patients with a history of vascular disease are at very high risk for acute coronary syndrome. Diabetes mellitus may result in atypical presentation of acute coronary syndrome. Dementia may impair a patient's ability to

---

EFFECT OF GENDER ON RISK OF ACUTE MYOCARDIAL INFARCTION AND ITS COMPLICATIONS

Although women suffer fewer acute myocardial infarctions than men, two factors argue for equal, if not greater, concern for women calling with chest pain. First, as with other risk factors, male gender as a risk factor refers to the general population, not an individual presenting with chest pain. In a given clinical situation of a patient presenting with chest pain, female gender does not lower the probability of acute myocardial infarction (10). Secondly, an acute myocardial infarction may be a more serious event in a woman, carrying a 38% 1-year mortality rate compared to 25% in men (2). The difference in short-term mortality is especially pronounced in younger women (again, a group usually considered "low risk"). As with the effect of age, the reason for this is multi-factorial, including delays and differences in treatment and co-morbidity (4). However, even adjusting for variations in medical history and treatment, the mortality rate in women is significantly higher. The implication is that women calling with chest pain should be managed as aggressively as men, if not more so.

describe symptoms. In conditions such as these, the physician should have a lower threshold for immediate in-person evaluation.

"Are You Taking Any Medications? Illicit Drugs?"

This question might provide clues to the patient's comorbidities such as those mentioned above. This is especially important if the patient does not seem to have a clear understanding of his or her medical history. Cocaine use greatly increases the risk of acute coronary syndrome.

"How Old Are You?"

As stated above, risk of complications from acute coronary syndrome increases with older age.

Referring the Patient to the Emergency Department for Acute Coronary Syndrome/Coronary Artery Disease

The patient should be discouraged from driving to the emergency department (either by himself or herself or by someone else) because this may cause delays in treatment. The patient should be instructed to call an ambulance so that emergency medical technicians can begin treatment immediately. The patient should also be instructed to chew one aspirin. If the patient has a previous history of coronary artery disease and has sublingual nitroglycerin available, he or she should take one dose immediately and repeat every 5 minutes for a total of three (4). However, this should not cause unnecessary delay if ED evaluation is indicated. If the first nitroglycerin is completely ineffective, the physician should consider activating emergency medical service while waiting for the second and third doses to be taken.

What Other Emergency Conditions Should Be Considered If the Patient Is Determined to Be at Low Risk for Acute Coronary Syndrome?

A large number of potentially life-threatening cardiac and pulmonary conditions also present with chest pain (Table 4-5). The clinical presentation of

---

Table 4-5   Other Potentially Life-Threatening Conditions That May Present with Chest Pain

- Dissecting aortic aneurysm
- Pericarditis
- Valvular disease
- Pneumonia
- Pneumothorax
- Pulmonary embolism

---

these conditions is well described in medical textbooks and is beyond the scope of this book, with the exception of pneumonia (see Chapter 7). However, few clinical studies examine the utility of the clinical history, either for telephone management or in-person management of these conditions.

HOW SHOULD CALLERS WITH LOW-RISK CHEST PAIN BE MANAGED?

If the physician decides that the patient's risk of acutely dangerous pathology is low, other diagnoses should be explored. Panic disorder, esophageal disorders, and muscular pain should be considered in patients who are unlikely to have a cardiac or pulmonary etiology for their chest pain (14). However, even if one of these conditions seems likely, patients should still be advised to call back if symptoms change and should probably be seen in person in an office setting within a few days to confirm the diagnosis.

Panic Attacks

Panic attacks have been shown to be present in one-third to one-half of patients presenting with chest pain (14,15). The diagnosis of panic attack is usually missed in the chest pain evaluation, resulting in inappropriate treatment and higher costs (15). A panic attack should be considered if the pain is associated with intense fear. Panic attacks typically come "out of the blue" and last a few minutes. Associated symptoms such as dyspnea and palpitations overlap with those of cardiac and pulmonary disease, so many patients will need to be evaluated in person before the diagnosis of panic disorder can be made.

Patients suspected of having panic attacks who are considered low risk for more urgent problems should be evaluated in person within a few days to assess for the utility of medications and psychotherapy. Deep breathing or other relaxation techniques can be recommended in the interim.

Esophageal Disorders

Esophageal disorders, especially gastroesophageal reflux and esophageal motility disorders, frequently cause chest pain (14). One study demonstrated an esophageal disorder in 75% of patients discharged from a critical care unit with undiagnosed chest pain (10). Heartburn, dysphagia, and odynophagia all suggest an esophageal disorder, but history alone, especially in the acute episode, may not be sufficient to determine a diagnosis. The discomfort of acute coronary syndrome may easily be confused with heartburn. As with panic attacks, many patients who are eventually diagnosed with esophageal chest pain will first need a cardiac evaluation (14). If the patient has no need for immediate in-person evaluation, antacids and behavioral measures can be recommended (see Chapter 5).

Muscular Pain

Muscular pain should be considered, especially if the pain is worse with palpation or position change. Over-the-counter analgesics such as acetaminophen, aspirin, or ibuprophen may be effective. Again, the patient should be seen in person within a few days to confirm the diagnosis.

WHAT TO TELL THE PATIENT

Patient Who Requires Emergency Evaluation

Provide the patient with the following information. After each point, verify the patient's understanding. After giving all the information, ask the patient to repeat what you have said back to you to confirm his or her understanding.

- "You may be having a heart attack [or other condition], which can be life threatening."
- "If this is a heart attack [or other condition], there are treatments that may prevent damage to your heart [or other complication] and may save your life."
- "You should go to the ED right away. The physicians there can do tests that can help us figure out exactly what is going on, and how we can help you."
- "You should call 911 now and tell them you are having chest pain; they will take you to the ED."
- "Don't drive to the ED. An ambulance is better because the emergency medical technicians may be able to start treating you right away."
- (For suspected acute coronary syndrome) "After you call 911, you should chew one aspirin. This will help keep your coronary arteries open."

Ask the patient to repeat the instructions and to tell you what he or she plans to do. If the patient is in severe distress, consider offering to call 911 for the patient. Driving a car to the ED (either by patient or family member) should be discouraged because studies have shown increased delays in treatment (8).

What to Do After You Hang Up

- *Call back soon:* If the patient seems ambivalent about calling 911, a call back in a few minutes might be very effective in conveying a sense of urgency.
- *Contact ED:* Making the ED aware of what you know about the patient may expedite immediate evaluation and treatment.
- *Contact the primary care physician:* Contacting the primary care provider either verbally or with a written note will help to ensure adequate follow-up.

What to Document

- Summary of symptoms
- What you told the patient regarding the risk of acute coronary syndrome or other condition and the need to call 911
- Whether patient verbalized understanding
- What the patient told you he or she would do

Patient Who Does Not Require Emergency Evaluation

- "Based on what you've told me, your chest pain is most likely not caused by a heart attack or other emergency problem."
- "The best thing to do now is to take it easy and avoid strenuous activity."
- "There is a small chance that this may indeed be a problem with your heart or something else serious. Call back right away if the pain changes at all, if you develop any new symptoms, or even if you're just worried."
- "You should see [me/your primary care physician] in the office within a few days to help figure out what's causing your symptoms."

After giving all the information, ask the patient to repeat it back to you to confirm his or her understanding. If another diagnosis is likely, you may mention that, but do not belittle the symptoms: you could discourage the patient from following up in the office or calling back if the symptoms change.

Medication and Other Treatment

- *If, based on the patient's symptoms, gastroesophageal reflux disease is likely*, consider antacids (see Chapter 5).
- *For probable musculoskeletal pain*, consider acetaminophen or non-steroidal anti-inflammatory medication.
- *For possible panic attacks*, consider advising the patient to take slow, deep breaths. Other relaxation techniques may be recommended depending on the physician's experience and

knowledge with these techniques. Although it is usually not advisable to initiate pharmacological treatment for panic disorder over the phone, the patient should be told that medication may help, and that he or she should be evaluated further.

What to Do After You Hang Up

- *Call back later:* If the patient is particularly concerned, if you think the symptoms may be evolving, or if you have any lingering concerns, calling the patient back is advised and often greatly appreciated by the patient. Ask about any changes in symptoms.
- *Contact the primary care physician:* A verbal or written communication will ensure adequate follow-up of patient's symptoms.

What to Document

- Summary of symptoms
- What the patient was told regarding low risk of acute coronary syndrome or other emergency
- When to call back
- When to follow-up with primary care provider
- Avoidance of strenuous activity
- Medication recommended (if applicable)
- Whether the patient verbalized understanding

HOW SHOULD NON-PHYSICIAN STAFF MANAGE CALLS FOR CHEST PAIN WHEN A PHYSICIAN IS NOT IMMEDIATELY AVAILABLE?

Without proper training, a well-intentioned but untrained staff member may cause needless delays in treatment when handling callers with chest pain. As this chapter has emphasized, acute coronary syndrome requires

PATIENT WHO CALLS WITH CHEST PAIN

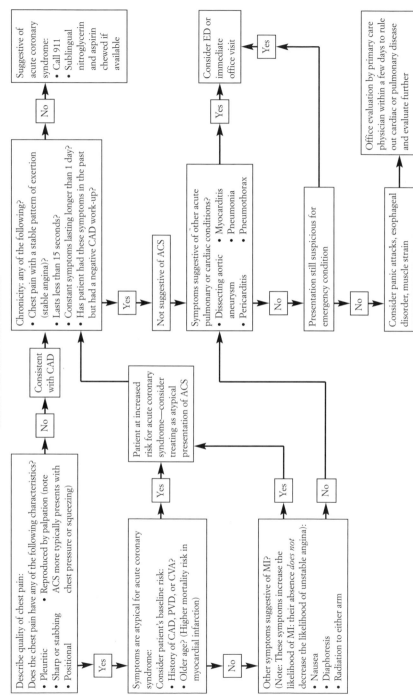

OFFICE STAFF EVALUATION OF PATIENT WHO CALLS WITH CHEST PAIN

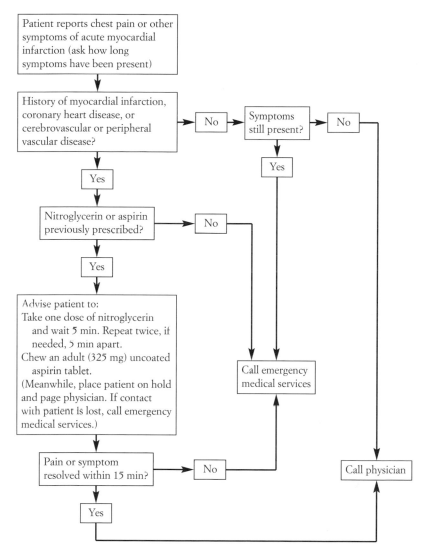

From Dracup K, Alonzo AA, Atkins JM, et al. The physician's role in minimizing prehospital delay in patients at high risk for acute myocardial infarction: recommendations from the National Heart Attack Alert Program. Ann Int Med. 1997;126:645–51; with permission.

immediate treatment to minimize the risk of death and myocardial injury. Studies examining causes for delays in treatment have found that delays are increased when patients with chest pain call their physicians, especially if the physician is not immediately available (8). The National Heart Attack Alert Program has published an algorithm and recommends that physicians train their staff members in its use (see page 103). Physicians may choose to simplify or otherwise modify this algorithm depending on the level of training of their staff. Regardless, a simple plan for management of callers with chest pain may have a significant impact on a caller's decision: most patients speak with someone before deciding to go to the ED, and even advice from a stranger will increase a patient's likelihood of going quickly to the ED (4).

REFERENCES

1. **Antman EM, Braunwald E.** Acute myocardial infarction. In: Braunwald E, Zipes DP, Libby P, eds. Braunwald's Heart Disease: A Textbook of Cardiovascular Medicine. Philadelphia: WB Saunders; 2001.
2. **Baigent C, Collins R, Appleby P, et al.** ISIS-2: 10 year survival among patients with suspected acute myocardial infarction in randomised comparison of intravenous streptokinase, oral aspirin, both, or neither. BMJ. 1998;316:1337-43.
3. American Heart Association Web page: www.americanheart.org/statistics/04comry.html. Based on the Atherosclerosis Risk in Communities (ARIC) Study of the National Heart, Lung, and Blood Institute (NHLBI), 1987-94.
4. **Vaccarino V, Parsons L, Every NR, et al.** Sex-based differences in early mortality after myocardial infarction. N Engl J Med. 1999;341:217-25.
5. **Lee TH, Goldman, L.** Primary care: evaluation of the patient with acute chest pain. N Engl J Med. 2000;342:1187-95.
6. **Hennekens CH, Albert CM, Godfried SL, et al.** Adjunctive drug therapy of acute myocardial infarction: evidence from clinical trials. N Engl J Med. 1996; 335:1660-7.
7. **Collins R, Peto R, Baigent C, Sleight P.** Aspirin, heparin and fibrinolytic therapy in suspected acute myocardial infarction. N Engl J Med. 1997;336:847-60.
8. **Dracup K, Alonzo AA, Atkins JM, et al.** The physician's role in minimizing prehospital delay in patients at high risk for acute myocardial infarction: recommendations from the National Heart Attack Alert Program. Ann Intern Med. 1997;126:645-1.
9. **Zalenski RJ, McCarren M, Roberts R, et al.** An evaluation of a chest pain diagnostic protocol to exclude acute cardiac ischemia in the emergency department. Arch Intern Med. 1997;157:1085-91.
10. **Panju AA, Hemmelgarn BR, Guyatt GH, et al.** Is this patient having a myocardial infarction? JAMA. 1998;280:1256-63.
11. **Lee TH, Cook EF, Weisberg M, et al.** Acute chest pain in the emergency room: identification and examination of low-risk patients. Arch Intern Med. 1985;145:65-9.

12. **Diamond GA, Forrester JS.** Analysis of probability as an aid in the diagnosis of coronary artery disease. N Engl J Med. 1979;300;1350-8.
13. **Paul SD, O'Gara PT, Majhoub ZA, et al.** Geriatric patients with acute myocardial infarction: cardiac risk factor profiles, presentation, thrombolysis, coronary interventions, and prognosis. Am Heart J. 1996;131:710-5.
14. **Katz PO.** Disorders of the esophagus: dysphasia, noncardiac chest pain, and gastroesophageal reflux. In: Barker LR, Burton JR, Zieve PD, eds. Principles of Ambulatory Medicine. Baltimore: Williams & Wilkins; 1999;459-70.
15. **Katerndahl DA, Trammell C.** Prevalence and recognition of panic states in STARNET patients presenting with chest pain. J Fam Pract. 1997;45:54-63.

5

ABDOMINAL PAIN

Ayse A. Atasoylu, MD, MPH

KEY POINTS

- The leading causes of abdominal pain vary with the population and the setting: among ambulatory patients, the leading causes are non-specific abdominal pain and peptic ulcer disease; in emergency department (ED) patients and hospitalized patients, non-specific abdominal pain and appendicitis; and among hospitalized elderly patients, infection (e.g., cholecystitis, urinary tract infection, diverticulitis), biliary tract disease, and intestinal obstruction.
- The telephone physician should have a lower threshold for in-person evaluation of elderly patients with any acute abdominal pain because older patients are more likely to present late in the course of their illness and to have a poor clinical outcome.
- Abdominal pain in the setting of significant bleeding, trauma, or recent abdominal surgery should prompt a referral to the emergency room without delay.
- The threshold for in-person emergency or urgent examination should be lowered in patients with acute abdominal pain who are taking narcotics, corticosteroids, anticoagulants, or nonsteroidal anti-inflammatory drugs (NSAIDs), for patients who are immunosuppressed or have significant co-morbid conditions; and for patients unable to take any food or fluid by mouth.

Continued

- Because the absence of warning signs will not rule out all acute serious conditions, the physician should consider common serious diagnoses even in the patient with no warning signs.
- Patients who call with abdominal pain in the absence of constipation and vomiting are not likely to have bowel obstruction.
- The diagnosis of acute appendicitis is rendered less likely by the absence of right lower quadrant pain and a history of similar previous pain. Patients with right lower quadrant pain or periumbilical pain that later localizes to the right lower quadrant should be seen in the ED.
- For patients in whom the disposition is not clear, the physician should consider calling back within a few hours to reassess the symptoms for resolution or progression.

Abdominal pain is one of the more challenging medical complaints to evaluate over the telephone. Because a delay in the diagnosis of some abdominal conditions, such as the surgical or acute abdomen, can potentially be fatal, the timeliness of an accurate diagnosis and appropriate management is essential for a favorable outcome (1). It is also important to identify which patients need non-emergency evaluation of their abdominal pain, as well as how to approach symptomatic management at home. This chapter outlines an approach to acute abdominal pain and chronic abdominal pain that has acutely changed in the adult patient. It is not intended to provide a comprehensive review of all abdominal conditions.

This chapter addresses the following central questions:

1. Does the patient have any warning signs for serious illness that should prompt an immediate visit to the ED?
2. If not, what diagnoses should be considered, and when should the patient be seen?
3. How should the patient with nonemergency abdominal pain be managed at home?

BACKGROUND

Epidemiology

Abdominal pain is a common reason for patients to call their general internist (2,3). It constitutes 40% of primary care office visits (4), accounts for 5% to 10% of all ED visits (5), and is the most common reason for hospital admission in the United States (6). In 1994, 5.3 million visits to the ED were for abdominal complaints (7).

The prevalence of abdominal pain diagnoses varies based on the population and setting being considered. In the ambulatory setting, non-specific abdominal pain and peptic ulcer disease comprise the leading diagnoses (8,9). A study looking at final diagnoses of patients examined in the ED found the following: non-specific abdominal pain (46%), appendicitis (16%), dyspepsia (8%), and cholecystitis (6%) (10). Another study looking at 1254 hospital patients found no specific cause of abdominal pain in 40% of patients on discharge, whereas appendicitis comprised 16.8% of the diagnoses (11).

In the elderly, the etiology and outcome of abdominal pain may be quite different than in younger groups. Various studies of elderly patients seen in the ED or admitted have identified common causes of abdominal pain as infection (including cholecystitis, urinary tract infection, and diverticulitis), mechanical intestinal obstruction, and biliary tract disease (12-14). Other common causes of abdominal pain in the elderly include constipation, malignancy, and kidney stones.

Utility of Early Diagnosis

Making an early diagnosis in the evaluation of abdominal pain is central to reducing morbidity and mortality. The progression to such outcomes as peritonitis, obstruction, sepsis, and death can be avoided by early recognition of a potentially serious etiology on the telephone, followed by prompt in-person assessment and treatment.

The mortality rate of appendicitis, for example, when diagnosed and treated in a timely manner, is less than 1% (5% to 15% in the elderly) (14,15). In contrast, mortality is significantly higher in the setting of appendiceal rupture. The incidence of perforation in patients with appendicitis ranges between 17% and 40% (60% to 70% in the elderly) (16,17). For acute cholecystitis, early diagnosis and treatment, leading to non-emergency surgery, can also help to decrease the occurrence of gallbladder perforation and death (3,18).

Furthermore, effective telephone assessment can help patients avoid unnecessary referral to the ED, the associated inconvenience, and hospital costs, not to mention morbidity from procedures that could otherwise have been avoided.

Early Diagnosis in the Elderly

Elderly patients are more likely to have underlying medical problems and tend to present later, resulting often in poorer clinical outcomes (19-22). Delayed presentation can occur because of chronic medical conditions such as diabetes, dementia, and cerebrovascular disease (20). Elderly patients admitted to the hospital with abdominal pain and chronic diseases have a higher mortality risk than those without chronic diseases (13% vs. 8%) (14,23). Other possible reasons for delayed presentation include a decrease in pain perception, nighttime symptoms attributed to indigestion, and difficulty in communicating an accurate medical history (19,20).

GENERAL APPROACH TO
THE TELEPHONE EVALUATION

The goal of telephone evaluation of abdominal pain is to ensure early detection of dangerous conditions and appropriate reassurance and home management of non-dangerous conditions. However, the wide variety of emergency conditions presenting with abdominal pain and the lack of studies on telephone evaluation of these conditions or the utility of the medical history in these conditions makes it difficult for a simple protocol to meet this goal.

When specific "warning signs" are present (see below), the decision to refer to the ED is straightforward. "Secondary warning signs" may also help guide the physician's decision making by lowering one's threshold for emergency or urgent (within 1 to 2 days) in-person evaluation.

Because the absence of all warning signs will not rule out all acutely dangerous conditions, the physician should consider the common dangerous diagnoses even in the patient with no warning signs. In such patients, the physician should consider other aspects of the history and presenting symptoms to rule out the common etiologies of serious abdominal pain based on the patient's age. For patients in whom the disposition is not clear, the physician should consider calling back within a few hours to reassess the symptoms for resolution or progression.

DOES THE PATIENT REQUIRE EMERGENCY EVALUATION?

Which Warning Signs Should Prompt a Visit to the Emergency Department?

Certain warning signs indicate a high risk for acutely dangerous pathology. Even if the remainder of the history sounds reassuring, the physician should strongly consider immediate ED referral (Table 5-1). These primary warning signs include bleeding (hematemesis, melena, profuse

---

Table 5-1   Primary and Secondary Warning Signs for Abdominal Pain

Primary Warning Signs (Should Prompt Immediate Referral to ED)

Bleeding (hematemesis, melena, profuse hematochezia)

Trauma

Pregnancy (consider referring to obstetrician)

Recent abdominal surgery or endoscopy

Abdominal pain in the setting of fever

Vomiting and unable to tolerate fluids

Patient unable to provide detailed and reliable history

Secondary Warning Signs (Should Lower the Threshold for Considering ED Evaluation)

Elderly patient

Patient taking narcotics, corticosteroids, anticoagulants, or NSAIDs

Patient with chronic medical conditions (diabetes, hepatitis, pancreatitis, inflammatory bowel disease, intra-abdominal malignancies, immunosuppression)

---

hematochezia), trauma, and recent abdominal surgery. Patients with abdominal pain and a recent endoscopic procedure require immediate evaluation and should also generally call the physician who performed the procedure, if available. Prompt evaluation in the ED is also warranted for any patient with abdominal pain unable to give a detailed and reliable medical history, patients with fever, and patients who have been vomiting and are unable to take anything by mouth and who may therefore need intravenous hydration. Patients with severe intractable abdominal pain without specific warning signs should probably be evaluated in the ED as well.

Abdominal pain in the setting of pregnancy may be difficult to fully assess over the telephone. Pregnant women with abdominal pain should be instructed to contact their obstetrician or go to the ED for further evaluation.

In some cases, based on the physician's judgment, patients with warning signs can be seen immediately within the office setting, rather than the ED. For example, the physician might opt to send a 26-year-old developmentally disabled man unable to give a detailed history with a history of constipation and abdominal pain to the office rather than the ED. A 40-year-old woman with no medical history, abdominal pain, and vomiting after a picnic, who is unable to keep down any food or water, might be evaluated promptly in the office if intravenous hydration can be administered there.

What Secondary Warning Signs Should Lower the Threshold for Emergency Evaluation?

The secondary warning signs listed in Table 5-1 may result in a higher risk of complications. Although these do not always necessitate emergency evaluation, they should lower the physician's threshold for emergency evaluation significantly. Symptoms of serious illness may be masked in patients on chronic corticosteroids or narcotics, or in those who are immunosuppressed. Special consideration should also be given to patients on anticoagulants or NSAIDs, because of the higher risk for bleeding, and the elderly because, as noted earlier, they are more likely to have underlying medical problems and tend to present later, which can result in poorer clinical outcomes. Referral to the ED should be strongly considered for patients with significant co-morbid conditions such as diabetes, hepatitis, pancreatitis, inflammatory bowel disease, or intra-abdominal malignancies.

IN A PATIENT WITHOUT WARNING SIGNS, WHAT DIAGNOSES SHOULD BE CONSIDERED, AND WHEN SHOULD THE PATIENT BE SEEN?

Unfortunately, the absence of all warning signs does not rule out all acutely dangerous conditions. Given that the list of common dangerous conditions is relatively short but includes conditions whose presentations vary greatly, a logical approach is to rule out these conditions systematically.

As mentioned above, the most common etiologies presenting as an acute abdomen include intestinal obstruction, acute appendicitis, and acute cholecystitis. The utility of elements of the history will be discussed briefly for each of these entities, as well as for ectopic pregnancy. Again, the focus of the telephone evaluation should be on asking about symptoms that will help rule out a diagnosis, rather than on trying to make a specific diagnosis. When a worrisome diagnosis cannot be ruled out, the patient should generally be referred for in-person examination. In some cases, the physician may opt to call the patient back within several hours to see if the pain has progressed or resolved before clarifying whether and when the patient should be examined in person.

COULD THE PATIENT HAVE A SMALL BOWEL OBSTRUCTION?

Small bowel obstruction is more common in patients over age 50 or with previous abdominal surgery (24). Additional aspects of the history that suggest this diagnosis include constipation or vomiting (11,24). In one study of patients with abdominal pain in the ED, no single variable could be used to rule out small bowel obstruction, but the absence of constipation and vomiting was useful in ruling out the diagnosis (negative predictive value 97%) (11). Conversely, the presence of both constipation and vomiting had a high specificity (95%) for small bowel obstruction.

On the telephone, the physician should identify those patients in whom small bowel obstruction cannot be ruled out, such as older patients with previous abdominal surgery. Such patients should be evaluated in person promptly. Patients with abdominal pain in the absence of constipation and vomiting are less likely to have obstruction; patients with abdominal pain with constipation and vomiting should be seen in person.

COULD THE PATIENT HAVE ACUTE APPENDICITIS?

For appendicitis, the incidence varies based on age and clinical setting. Between 12% and 26% of patients who present to an ED with acute abdominal pain of less than 1 week's duration have acute appendicitis (10,25-27). Twenty-five percent of patients younger than 60 years of age evaluated for acute abdominal pain in EDs or surgical services have acute appendicitis compared to an incidence of 4% in patients over the age of 60 years (9,14,15,28-30). In ambulatory patients with abdominal pain, the prevalence is much lower, ranging between 0.7% and 1.6% (4,9).

One review examined the relative importance of individual signs and symptoms in the diagnosis of acute appendicitis among ED or surgical patients presenting with abdominal pain of less than 1 week's duration (31). No single component of the medical history could rule out appendicitis; however, the presence of similar previous pain and the absence of right lower quadrant pain made the diagnosis less likely (negative likelihood ratios 0.3 and 0.2, respectively). Likewise, the absence of the classic migration of pain (from the periumbilical area to the right lower quadrant) significantly reduced the likelihood of acute appendicitis (negative likelihood ratio 0.5) (31) (Table 5-2).

The presence of right lower quadrant pain and the migration of initial periumbilical pain to the right lower quadrant were identified as the most useful elements of the medical history to rule in acute appendicitis (positive likelihood ratios 8.0 and 3.1, respectively) (31).

When assessing the possibility of acute appendicitis over the phone, the absence of right lower quadrant pain, the absence of the classic migration of pain, and the presence of similar previous pain makes the diagnosis less likely, whereas the presence of right lower quadrant pain or a history of periumbilical pain that later localizes to the right lower quadrant should raise a strong suspicion for the diagnosis; such patients should be seen in the ED. Abdominal pain associated with anorexia, nausea, and/or vomiting

Table 5-2   Likelihood Ratios of Individual Symptoms in the Diagnosis of Acute Appendicitis

Symptom	*Positive Likelihood Ratio (LR+)*
Right lower quadrant pain	8.0
Migration of periumbilical pain to the right lower quadrant	3.1
	Negative Likelihood Ratio (LR–)
Absence of right lower quadrant pain	0.2
Presence of similar previous pain	0.3
Absence of migration of pain from the peri- umbilical area to the right lower quadrant	0.5

Modified from Wagner JM, McKinney WP, Carpenter JL. Does this patient have appendicitis? JAMA. 1996; 276:1589-94.

may also occur with appendicitis, but the absence of these factors does not permit the diagnosis to be ruled out.

COULD THE PATIENT HAVE ACUTE CHOLECYSTITIS?

An estimated 20% of adult Americans have gallstones, and 30% of these people will at some point develop acute cholecystitis (32). Approximately 30% of patients presenting with classic symptoms of acute cholecystitis (fever with sudden unremitting epigastric or right upper quadrant pain often radiating to the right scapula) have the disease (33). The incidence of acute cholecystitis increases with age (10). Most patients will have a history of recurrent episodes of biliary colic. Other patients may have non-pain symptoms such as gaseousness, bloating, indigestion, and fatty food intolerance (34). Elderly patients with acute cholecystitis frequently present without classic symptoms (35).

One prospective cohort study of patients admitted with acute abdominal pain identified some elements of the medical history that may increase the likelihood of acute cholecystitis (36). These included pain in the right upper quadrant, previous history of pain in this area, and the presence of either jaundice or vomiting. No criteria were identified that could reliably exclude the diagnosis. In the presence of such symptoms, acute cholecystitis should be considered, and the patient should be seen in the office or ED for prompt evaluation and imaging studies. When dealing with elderly patients, even those with less typical symptoms, the physician should have an even lower threshold for prompt in-person evaluation.

Patients who call with symptoms suggestive of acute cholecystitis should be referred for ED or office evaluation. There is no evidence that the absence of particular symptoms helps to rule out the diagnosis. In some cases, the physician may opt to call the patient back in several hours to assess the progression or resolution of the pain before making a decision about whether or when the patient should be seen.

COULD THE PATIENT HAVE ECTOPIC PREGNANCY?

A recent prospective observational study of female patients presenting to an urban ED with abdominal pain or vaginal bleeding and a positive beta-HCG identified factors associated with an increased risk of ectopic pregnancy (37). Moderate to severe pain, sharp pain, and the lateral location of pain were all associated with a higher risk of ectopic pregnancy, whereas the midline location of pain decreased the risk. Factors found not to be statistically significant predictors of ectopic pregnancy included previous ectopic pregnancy and a history of pelvic inflammatory disease, no bleeding or mild bleeding, and the passage of tissue (38). No combination of findings was identified that could rule in or rule out the diagnosis reliably (37,38). Therefore, all cases of possible ectopic pregnancy (e.g., a sexually active woman with abdominal pain not using reliable contraception or whose menstrual period is late) should be referred for prompt in-person evaluation to rule out the diagnosis (38).

COULD THE ACUTE ABDOMINAL PAIN HAVE ANOTHER CAUSE?

Other causes of an acute abdomen, such as diverticulitis, perforated abdominal viscus, pelvic inflammatory disease, and nephrolithiasis, should be considered when symptoms are suggestive. A review of the literature produced no studies focusing on the history that would be applicable to the telephone evaluation of any of these diagnoses.

HOW SHOULD THE PATIENT WITH NONEMERGENCY ABDOMINAL PAIN BE MANAGED?

The most common causes of non-emergency abdominal pain include peptic ulcer disease, biliary colic, and abdominal pain that is non-specific in nature. There is a paucity of studies focusing on the sensitivity and specificity of symptoms of non-emergency abdominal pain, but once the more serious acute causes are ruled out using the criteria above, the history can be helpful in guiding recommendations until the patient is seen in the office.

Non-Specific Abdominal Pain

For non-specific abdominal pain without symptoms of more serious illness or warning signs, it may be appropriate to see the patient in person within the next day or two. As mentioned above, some dangerous conditions may present non-specifically, especially in the elderly.

DYSPEPSIA

Patients calling with symptoms of dyspepsia may describe epigastric pain or discomfort, bloating, nausea, or belching. Peptic ulcer disease and gastroesophageal reflux are two of the more common causes of dyspepsia; other causes include irritable bowel syndrome and less commonly esophagitis, cholelithiasis, gastritis, and duodenitis.

For patients calling with symptoms of dyspepsia without any warning signs or symptoms of more serious disease, it is usually appropriate to prescribe symptomatic treatment over the telephone. Patients can be instructed to take an antacid (such as calcium carbonate or magnesium carbonate) or a histamine-2 antagonist with instructions to avoid

PATIENT WHO CALLS WITH ABDOMINAL PAIN

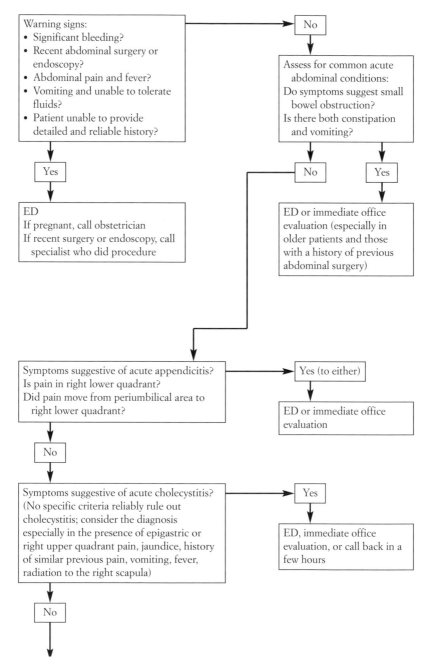

Warning signs:
- Significant bleeding?
- Recent abdominal surgery or endoscopy?
- Abdominal pain and fever?
- Vomiting and unable to tolerate fluids?
- Patient unable to provide detailed and reliable history?

No

Assess for common acute abdominal conditions:
Do symptoms suggest small bowel obstruction?
Is there both constipation and vomiting?

Yes

No

Yes

ED
If pregnant, call obstetrician
If recent surgery or endoscopy, call specialist who did procedure

ED or immediate office evaluation (especially in older patients and those with a history of previous abdominal surgery)

Symptoms suggestive of acute appendicitis?
Is pain in right lower quadrant?
Did pain move from periumbilical area to right lower quadrant?

Yes (to either)

ED or immediate office evaluation

No

Symptoms suggestive of acute cholecystitis? (No specific criteria reliably rule out cholecystitis; consider the diagnosis especially in the presence of epigastric or right upper quadrant pain, jaundice, history of similar previous pain, vomiting, fever, radiation to the right scapula)

Yes

ED, immediate office evaluation, or call back in a few hours

No

Cont'd on page 119

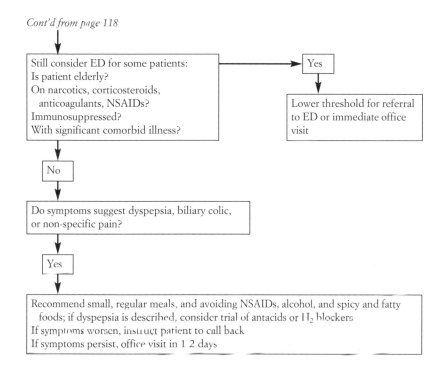

Cont'd from page 118

Still consider ED for some patients:
Is patient elderly?
On narcotics, corticosteroids,
  anticoagulants, NSAIDs?
Immunosuppressed?
With significant comorbid illness?

Yes

Lower threshold for referral to ED or immediate office visit

No

Do symptoms suggest dyspepsia, biliary colic, or non-specific pain?

Yes

Recommend small, regular meals, and avoiding NSAIDs, alcohol, and spicy and fatty foods; if dyspepsia is described, consider trial of antacids or H₂ blockers
If symptoms worsen, instruct patient to call back
If symptoms persist, office visit in 1-2 days

non-steroidal anti-inflammatory medications. Follow-up in the office should be arranged within 2 days. Patients who develop warning signs or worsening pain should have an in-person assessment sooner than 2 days.

BILIARY COLIC

Symptoms of biliary colic have been characteristically described as the sudden onset of steady severe aching pain in the right upper quadrant or epigastrium that waxes and wanes; is aggravated by fatty foods; frequently radiates to the right shoulder, scapula, or interscapular area; and is occasionally associated with nausea and vomiting. However, studies suggest that there is a lack of precision in defining the symptoms related to diseases of the gallbladder (34).

If uncomplicated biliary colic is suspected, the patient should be instructed to avoid fatty foods and to follow up in the office within 2 days.

Patients who develop warning signs or worsening pain should have an in-person assessment sooner than 2 days.

WHAT TO TELL THE PATIENT

Patient Requires Emergency Evaluation

- Urge the patient to go to the ED.
- Explain that the symptoms sound serious and warrant an examination; tell the patient which warning signs are present, if any, and why they can be dangerous.
- Ensure that the patient has a ride to the ED, or that he or she should call 911 if appropriate.

Patient Does Not Require Emergency Evaluation

- Explain that the type of pain does not sound as if it needs immediate ED evaluation.
- Describe warning signs that should prompt an immediate call back or ED visit (e.g., black tarry stools, rectal bleeding, bloody vomiting, fevers, inability to keep down fluids, or severely worsening pain.)
- Explain that a diagnosis is difficult without an in-person examination.
- Recommend eating small, regular meals, and avoiding NSAIDs, alcohol, and spicy and fatty foods. If dyspepsia is described, recommend trying an antacid.
- Urge the patient to call back if the pain worsens.
- Urge the patient to make an appointment within the next day or two if the pain continues.

WHAT TO DOCUMENT

Document the chief complaint in the patient's words, the presence or absence of warning signs, your working diagnosis, instructions to the patient and his or her understanding of your instructions, and the follow-up plan.

REFERENCES

1. **Rogers AI.** The acute abdomen. Compr Ther. 1998;4:25-33.
2. **Peterson MC, Holbrook JH, Hales DV, et al.** Contributions of the history, physical examination and laboratory investigation in making medical diagnoses. West J Med. 1992;156:163-5.
3. **Johnson LB.** The importance of early diagnosis of acute acalculus cholecystitis. Surg Gynecol Obstet. 1987;164:197-203.
4. **Britt H, Bridges-Webb C, Sayer GP.** The diagnostic difficulties of abdominal pain. Aust Fam Physician. 1994;23:375-81.
5. **American College of Emergency Physicians.** Clinical policy for the initial approach to patients presenting with a chief complaint of non-traumatic acute abdominal pain. Ann Emerg Med. 1994;23:906-22.
6. **McFadden DW, Zinner MJ.** Approach to the Patient with the Acute Abdomen and Fever of Abdominal Origin. Philadelphia: JB Lippincott; 1991.
7. **Stussman BJ.** National Hospital Ambulatory Medical Care Survey. Vital and Health Statistics of the Centers for Disease Control and Prevention/National Center for Health Statistics. 1996;275:1-17.
8. **Klinkman MS.** Episodes of care for abdominal pain in a primary care practice. Arch Fam Med. 1996;5:279-85.
9. **Wasson JH, Sox IIC, Sox CH.** The diagnosis of abdominal pain in ambulatory male patients. Med Decis Making. 1981;1:215-24.
10. **Wilson DH.** The acute abdomen in the accident and emergency department. Practitioner. 1979;220:440-5.
11. **Bohner H, Yang Q, Franke C, et al.** Simple data from history and physical examination help to exclude bowel obstruction and to avoid radiographic studies in patients with acute abdominal pain. Eur J Surg. 1998;164:777-84.
12. **Marco CA, Schoenfeld CN, Keyl PM, et al.** Abdominal pain in geriatric emergency patients: variables associated with adverse outcomes. Acad Emerg Med. 1998;5:1163-8.
13. **Ponka JL, Welborn JK, Brush BE.** Acute abdominal pain in aged patients. An analysis of 200 cases. J Am Geriatr Soc. 1983;11:993.
14. **Fenyo G.** Acute abdominal disease in the elderly. Am J Surg. 1982;143:751-4.
15. **Balsano N, Cayten CG.** Surgical emergencies of the abdomen. Emerg Med Clin North Am. 1990;8:399-410.
16. **Lewis FR, Holcroft JW, Boey J, Dunphy E.** Appendicitis: a critical review of diagnosis and treatment in 1000 cases. Arch Surg. 1975;110:677-84.
17. **Addiss DG, Shaffer N, Fowler BS, Tauxe RV.** The epidemiology of appendicitis and appendectomy in the United States. Am J Epidemiol. 1990;132:910-25.
18. **Van der Ham AC, Lange JF, Yo TI.** Acute cholecystitis in the elderly: a retrospective study. Neth J Surg. 1986;38:142-6.
19. **Bender JS.** Approach to the acute abdomen. Geriatr Med Clin North Am. 1989;73:1413-22.
20. **Sanson TG, O'Keefe KP.** Evaluation of abdominal pain in the elderly. Gastrointestinal Emergencies, Part 1. Emerg Med Clin North Am. 1996;13:615-27.

21. **Rothrock SG, Greenfield RH.** Acute abdominal pain in the elderly: clues to identifying serious illness. Part 1: Clinical assessment and diagnostic studies. Emergency Medicine Reports. 1992;13:177-84.
22. **Rothrock SG, Greenfield RH.** Acute abdominal pain in the elderly: clues to identifying serious illness: Part 2—Diagnosis and management of common disorders. Emergency Medicine Reports. 1992;13:177-84.
23. **Kizer KW, Vassar MJ.** Emergency department diagnosis of abdominal disorders in the elderly. Am J Emerg Med. 1998;16:357-62.
24. **Eskelinen M, Ikonen J, Lipponen P.** Contributions of history-taking, physical examination, and computer assistance to diagnosis of acute small-bowel obstruction: a prospective study of 1,333 patients with acute abdominal pain. Scand J Gastroenterol. 1994;29:715-21.
25. **Adams ID, Chan M, Clifford PC, et al.** Computer aided diagnosis of acute abdominal pain: a multicenter study. BMJ.1986;293:800-04.
26. **De Dombal ET, Leaper DJ, Horrocks JC, et al.** Human and computer-aided diagnosis of abdominal pain: further reports with emphasis of performance of clinicians. BMJ. 1994;1:376-80.
27. **De Dombal ET.** Educational assessment of clinical diagnostic skills: studies across Europe on acute abdominal pain. Postgrad Med J. 1993;69:S94-S96.
28. **Irvin TT.** Abdominal pain: a surgical audit of 1190 emergency admissions. Br J Surg. 1989;76:1121-5.
29. **Bugiliosi TF, Meloy TD, Vukov LF.** Acute abdominal pain in the elderly. Ann Emerg Med. 1990;19:1383-6.
30. **Fenyo G.** Diagnostic problems of acute abdominal disease in the aged. Acta Chir Scand. 1994;140:396-405.
31. **Wagner JM, McKinney WP, Carpenter JL.** Does this patient have appendicitis? JAMA. 1996;276:1589-94.
32. **Cox GR, Browne BJ.** Acute cholecystitis in the emergency department. J Emerg Med. 1989;7:501-11.
33. **Black ER, Bordley DR, Tape TG, Panzer RJ.** Diagnostic Strategies for Common Medical Problems. Philadelphia: American College of Physicians; 1999:158-69.
34. **Fredrick LF, Lonborg R, Thirlby R, et al.** What symptoms does cholecystectomy cure? Insights from an outcome measurement project and review of the literature. Am J Surg. 1995;169:533-8.
35. **Parker LJ, Vukov LF, Wollan PC.** Emergency department of geriatric patients with acute cholecystitis. Acad Emerg Med. 1997;4:51-5.
36. **Eskelinen M, Ikonen J, Lipponen P.** Diagnostic approaches in acute cholecystitis: a prospective study of 1,333 patients with acute abdominal pain. Theor Surg. 1993;8:15-20.
37. **Dart RG, Kaplan B, Varaklis K.** Predictive value of history and physical examination in patients with suspected ectopic pregnancy. Ann Emerg Med. 1999;33:283-90.
38. **Gutman SJ, Lindsay K.** Suspected ectopic pregnancy. Can it be predicted by history and examination? Can Fam Physician. 2000;46:1297-8.

6

DIARRHEA

Ira Daniel Breite, MD • Sondra Zabar, MD

KEY POINTS

- Most episodes of acute diarrhea are self-limited and do not require or result in physician assistance.
- Morbidity and mortality from diarrhea are primarily a result of dehydration and electrolyte losses.
- Physicians dealing with a patient with diarrhea on the telephone should separate patients into one of four groups based on need for further evaluation:
    1. *Low risk:* needs no further follow-up
    2. *Low risk acutely, possible underlying pathology:* needs follow-up non-urgently for possible neoplasm or colitis
    3. *Moderate risk:* needs to be seen urgently (within 1 to 2 days) and followed closely with repeat calls during the first 24 hours to ensure response to home treatment
    4. *High risk:* emergency in-person evaluation and treatment, either in the emergency department or in a well-equipped physician's office.
- Warning signs of acutely dangerous diarrheal illness reflect evidence of severe volume depletion, an invasive pathogen, or being in a high-risk population. These warning signs include:
    1. Profuse watery diarrhea with dehydration
    2. Passage of six or more stools in a 24-hour period or diarrhea for more than 2 days

Continued

3. Passage of many small volume stools with blood and mucous
4. Temperature greater than 38.5°C (101°F)
5. Severe abdominal pain
6. Patient who is elderly, immunocompromised, or has significant comorbidity, such as severe coronary artery disease.

- When no warning signs are present, the acute illness can be managed at home with fluids and anti-diarrheal medication.
- When warning signs are present, the patient should be evaluated as an emergency in person or be followed closely on the phone and given the soonest possible office appointment.
- Patients who are otherwise healthy, not elderly, can drink and keep down fluids, and can self-monitor for worsening symptoms can often be treated at home with telephone follow-up until the next possible office appointment.
- Patients with warning signs should drink fluids that contain electrolytes. When there are no warning signs, water and saltines or other starch with salt will usually suffice.
- Bismuth products such as Pepto-Bismol are generally safe and effective in reducing stool frequency and nausea.

Although most patients with acute diarrhea improve with symptomatic treatment at home (1,2), the physician needs to identify patients at high risk for complications such as severe dehydration, sepsis, and death. Physicians must also identify those requiring non-urgent referral for evaluation for underlying problems, such as colon cancer or inflammatory bowel disease. Finally, aggressive home treatment with the proper medications and appropriate fluids will lead to rapid symptomatic improvement in many patients and may prevent progression to severe dehydration.

This chapter focuses on acute diarrhea. Diarrhea may be defined as three or more loose stools a day or two loose stools a day with other abdominal symptoms. Diarrhea that lasts more than 14 days is usually classified as persistent. Diarrhea that lasts more than 1 month is defined as chronic (3). Patients calling with persistent or chronic symptoms that have not changed acutely should be reassured that they have a low risk of acute complications, but should make an appointment to see their primary care physician in the office in a timely manner.

Acute diarrhea is rarely a life-threatening illness. The first task for the telephone physician is to identify those few patients that need to be evaluated in person, determine how soon they should be seen, and where. Most patients will benefit from advice on symptomatic treatment. This chapter therefore focuses on the following central questions:

1. What elements of the history suggest that the patient may require prompt in-person evaluation for severe dehydration and/or severe infection?
2. What factors indicate that a patient is high risk, requiring emergency in-person evaluation (immediately), as opposed to moderate risk, requiring urgent evaluation (within a few days)?
3. For patients requiring emergency in-person evaluation, what factors indicate that a patient should be seen in an emergency department, as opposed to a physician's office?
4. Which patients require non-urgent evaluation for possible significant underlying pathology?
5. How should patients not requiring urgent evaluation be managed?

BACKGROUND

Epidemiology

Diarrhea is an extremely common condition. It is estimated that each American adult will contract one case of acute diarrhea every year (1,2). Half of these patients have restriction of their activities because of the diarrhea. Only one in 12 with diarrhea seek medical treatment and only 3% of those are hospitalized. The vast majority of cases of acute diarrhea resolve spontaneously.

Utility of Early Telephone Evaluation

Regardless of the etiology, the main direct causes of morbidity and mortality are usually dehydration and electrolyte derangements (4). Telephone evaluation early in the course of the illness may prevent dehydration and electrolyte derangements by ensuring appropriate fluid intake and curtailing further fluid losses with anti-diarrhea medication. Telephone consultation may also provide benefit from more rapid symptomatic improvement and avoidance of an emergency department visit.

Physicians can identify patients who are at high risk for complications and advise them on the most appropriate setting and timing for in-person evaluation based on their risk. The physician can also identify patients who are at risk for conditions such as neoplasms or inflammatory bowel disease that require further testing, but may not be true emergencies.

TRAVELER'S DIARRHEA

Travel to certain countries and regions entails a 30% chance of infectious diarrhea. Because it has a different epidemiology, traveler's diarrhea is approached differently. Endemic regions are Mexico, many Asian and African countries, mountainous areas, or recreational bodies of water in North America. Pathogens are 80% bacterial, and the most common is enterotoxigenic *E. coli* (10). Its presentation ranges from mild, self-limited watery diarrhea to acute dysentery with fever and bloody stools. Symptoms may not begin for 5 to 15 days, so patients may not think to mention they have just returned from a trip. Patients may also call while in the endemic area. Treatment involves hydration and medication. Mild diarrhea usually responds to bismuth subsalicylate (6). More severe cases may require an antibiotic such as a quinolone. Antibiotics usually lead to relief of symptoms within 1 day (11). Traveler's diarrhea is an exception to the rule of "never prescribe antibiotics for diarrhea without a culture." In patients who present with probable traveler's diarrhea, it is reasonable to prescribe either ciprofloxacin 500 mg bid or norfloxacin 400 mg bid for 1 to 2 days. Most patients will respond to the above treatment and do not need further evaluation. If the diarrhea persists despite quinolone treatment, the patient should be evaluated for parasitic disease.

GENERAL APPROACH ON THE TELEPHONE

The primary goal of managing acute diarrhea over the telephone is the prevention of severe dehydration and relief of symptoms. There are no clinical studies examining the management of diarrhea over the telephone. There are, however, a number of relevant publications including clinical trials on outpatients and expert recommendations such as the American College of Gastroenterology Guidelines on Acute Infectious Diarrhea in Adults (3). The following approach seeks to adapt these guidelines and other studies for the specific needs of the telephone physician.

Patients calling for acute diarrhea can be classified according to their risk for complications or underlying pathology: 1) low risk, 2) low risk acutely, possible underlying pathology, 3) moderate risk acutely, 4) high risk. These classifications reflect a focus on management (see Table 6-1) to avoid complications and not on establishing a specific etiology.

Determination of a specific etiologic agent on the telephone is impractical and unnecessary. Even in patients undergoing full laboratory evaluation, only about 2% of stool exams for bacterial culture or ova and parasites demonstrate a specific etiology (5). Whether from a viral source or as a result of food-borne illness, most cases will be self-limited and have similar constellations of symptoms.

Table 6-1 Classification and Management of Patients Calling with Diarrhea

Group	Treatment
Low risk	Home treatment with fluid and anti-diarrheal medications
Low risk (acute, possible underlying pathology)	Home treatment with fluid and anti-diarrheal medications; nonurgent referral for diagnostic evaluation
Moderate risk	Home treatment with fluid and anti-diarrheal medications; follow closely with repeat calls over 24 hr; urgent office visit (within 1–2 days)
High risk	Emergency medical evaluation in office or ED

WARNING SIGNS: WHAT ELEMENTS OF THE HISTORY SUGGEST THAT THE PATIENT MAY REQUIRE PROMPT IN-PERSON EVALUATION FOR SEVERE DEHYDRATION AND/OR SEVERE INFECTION?

Three types of situations should prompt the physician to advise in-person evaluation urgently or as an emergency: 1) the patient is in a high-risk group, such as the elderly, 2) severe dehydration, and 3) probable invasive pathogen that may require stool studies and antibiotics (see Table 6-2) (3,6).

A patient with one or more of these signs should be considered to be at moderate or high risk for acute complications (see the section below for distinguishing high-risk from moderate-risk patients). The physician should strongly consider having the patient evaluated in-person (in office or emergency department) urgently or as an emergency. At a minimum, physicians should provide close follow-up on the phone for the next 24 hours.

When none of these warning signs are present, the patient is probably at low risk for acute complications from diarrhea, and can usually be managed with home treatment. Given the importance of these warning signs to the triage decision, the telephone interview should initially focus on ascertaining whether these warning signs are present.

Is the Patient Dehydrated?

The first two warning signs in Table 6-2 concern dehydration or the possibility of dehydration. Determining volume status over the phone can be a challenge. Mild dehydration in an otherwise healthy patient is usually not cause for much concern, as long as the patient can drink fluids well. On the other hand, some patients may tolerate even mild dehydration poorly because of their comorbidities. Patients with conditions such as coronary artery disease or pregnancy should be seen promptly regardless of their apparent volume status (see discussion of warning signs and Table 6-2).

Because mild dehydration is not usually serious in otherwise healthy patients, the physician may choose to focus on determining the presence of more severe volume depletion. The most useful symptom for assessing

Table 6-2 Warning Signs of Risk for Complications from Diarrhea

Warning Sign	*Possible Implication of Warning Sign*
Profuse watery diarrhea with dehydration	Hypovolemic shock
Passage of 6 or more stools in a 24-hour period *or* diarrhea for more than 2 days	Greater risk for severe dehydration
Passage of many small-volume stools with blood and mucus	Invasive infection (e.g., *Shigella*, entero-hemorrhagic *E. coli* [EHEC]); disease may progress without antibiotics
Temperature greater than 38.5°C	Same as for blood and mucus: invasive bacteria, *C. difficile; E. histolytica;* may also be enteric viruses or inflammatory bowel disease; perforation, a rare complication of diarrhea, can occur in these situations
Severe abdominal pain (mild, crampy pain is not a warning sign)	Ischemic colitis, diverticulitis, partial SBO (3), bacterial pathogens (6)
High-risk group: elderly, immunocompromised, other significant comorbidity (e.g., severe coronary artery disease)	*Elderly:* increased risk of mortality from acute diarrhea; *immunocompromised:* more complicated course and more unusual pathogens; *other comorbidity:* increased risk of complication

Data from References 3 and 6.

this is severe postural dizziness. This symptom is highly predictive for severe volume loss, with a sensitivity of 97% (95% confidence interval [CI]: 91%-100%) and a specificity of 98% (95% CI: 97%-99%) (7). No evidence supports mild postural dizziness as a predictor of volume depletion.

Invasive Versus Noninvasive Diarrhea

Although identification of a specific organism causing acute diarrhea is usually not necessary, it is useful to determine whether the diarrhea is likely to be invasive or noninvasive because there are significant implications for management (6). Table 6-3 compares the important features of these two types.

Invasive diarrhea typically presents with frequent small stools that may be bloody and/or mucoid. Fever, abdominal pain, and tenesmus may also be present. Causes include *Shigella*, enteroinvasive *E. coli*, and

Table 6-3 Comparison of Invasive and Noninvasive Diarrhea

	Invasive Diarrhea	*Noninvasive Diarrhea*
Prevalence in acute diarrhea	10%	90%
Character of stools	Frequent small stools— bloody or mucoid	Frequent watery stools, may be very large volume
Associated symptoms	Fever, abdominal pain, tenesmus	Absence of fever, abdominal pain, or tenesmus
Common causes	*Shigella*, enteroinvasive *E. coli*, enterohemorrhagic *E. coli* (EHEC)	Viruses, pre-formed toxin ingestion (such as *S. aureus*)
Utility of work-up for etiologic agent	Treatment for specific etiology may reduce complications	Not generally helpful
Complications	Dehydration, (rarely) sepsis	Dehydration and complications thereof
Treatment	Fluids and electrolyte replacement, possibly antibiotics	Fluid and electrolyte replacement, anti-diarrheal medication

Data from Reference 6.

enterohemorrhagic *E. coli*. Noninvasive diarrhea is much more common (90% of acute diarrheal illnesses) and presents with watery diarrhea without the above symptoms (6) and is most commonly caused by viruses. Although invasive diarrhea should generally be evaluated promptly for a specific diagnosis and treatment, noninvasive diarrhea can usually be treated with fluid replacement and anti-diarrhea medication.

WHAT FACTORS INDICATE THAT A PATIENT IS HIGH RISK, REQUIRING EMERGENCY IN-PERSON EVALUATION (IMMEDIATELY), AS OPPOSED TO MODERATE RISK, REQUIRING URGENT EVALUATION (WITHIN A FEW DAYS)?

The decision to advise immediate in-person evaluation (either in an emergency department or in the physician's office), as opposed to advising office

---

Table 6-4 Conditions to Consider When Deciding Between Emergency and Urgent In-Person Evaluation

- Patient sounds "toxic" versus comfortable
- Impaired cognitive ability (either acute or chronic)
- Patient lives alone
- Unable to drink fluids
- Severe comorbidity (severe coronary disease, recent stroke, pregnancy)
- Time until next available office appointment
- Patient preference

---

evaluation within a day or two, is largely a judgement call based on a thorough knowledge of the particular clinical situation. No studies support using any specific criteria for this decision, yet a number of clinical situations logically indicate a greater degree of risk or otherwise makes an emergency evaluation preferable (Table 6-4). Physicians should inquire about these conditions, especially in situations when an immediate office visit is not possible. Patients can usually wait a day or two for an office visit if they are otherwise healthy, can drink fluids, and can self-monitor for worsening symptoms. Assistance at home to help them with their treatment also makes deferring in-person evaluation more reasonable. These patients, or their companions, can always call back or go to the emergency department if their symptoms worsen.

On the other hand, patients should be seen immediately if they cannot drink fluids or have an impaired ability to self-monitor. Patients should also be considered high-risk if they are elderly, immunocompromised, or have comorbidities that put them at high risk for other problems such as coronary or cerebral vascular events.

If it is decided that a patient with warning signs may wait a day or two before in-person evaluation (urgent evaluation), the physician should make plans to follow up with the patient over the phone within the first few hours and then daily to ensure the illness does not worsen. Either the physician or the patient can initiate these calls, but either way a schedule for calls should be agreed upon (see below, "What to Tell the Patient").

This then raises the question: if the patient gets better while waiting for the appointment, is in-person evaluation still necessary? This is a clinical judgement, based on how ill the patient was initially, how much he or she has improved, and what his or her general health is. Patients with rectal bleeding should be seen regardless of improvement, although the visit need not be urgent (see the section below discussing need for Non-Urgent Evaluation).

FOR PATIENTS REQUIRING EMERGENCY IN-PERSON EVALUATION, WHAT FACTORS INDICATE THAT A PATIENT SHOULD BE SEEN IN AN EMERGENCY DEPARTMENT RATHER THAN IN A PHYSICIAN'S OFFICE?

Some physicians' offices are well suited for the care of these acutely ill patients, whereas some clearly are not: the physician should consider what his or her options would be should the patient be found in need of immediate fluid resuscitation. Similarly, not all patients will have a support system of family or friends in place to help them get to and from the physician's office despite an obviously disruptive illness. Other patients will just sound too ill to do anything but go to an emergency department. Table 6-5 lists factors that should be taken into account when considering this decision.

Table 6-5 Considerations When Deciding Between Immediate Office Evaluation and Emergency Department Referral

---

- Patient seems too sick, weak, or uncomfortable to get to the office
- Availability of companion to help patient get to office
- Patient preference
- Proximity of physician's office to hospital
- Availability of lab tests and intravenous therapy in physician's office
- Availability of office appointments that day

DOES THE PATIENT REQUIRE NON-URGENT EVALUATION (WITHIN ONE WEEK) FOR POSSIBLE SIGNIFICANT UNDERLYING PATHOLOGY?

A number of non-acute diseases may present with an episode of acute diarrhea. Although the diarrhea may be mild and self-limited, the underlying disease might be more significant, such as colorectal cancer and inflammatory bowel disease. Patients with rectal bleeding or multiple episodes of diarrhea should be evaluated non-urgently (within a week) in person and will almost certainly need to be referred for endoscopic evaluation. Also, patients with chronic or persistent diarrhea should be evaluated for a diagnosis non-urgently.

HOW SHOULD PATIENTS NOT NEEDING URGENT IN-PERSON EVALUATION BE MANAGED?

As stated above, most patients with acute diarrhea do not need urgent in-person evaluation. These patients should be instructed on self-management with oral fluids and anti-diarrheal medications and given specific instructions when to call back. Patients should be instructed to stop any unnecessary medications that might be causing diarrhea (Table 6-6).

Table 6-6   Medications That May Cause Diarrhea

- Most antibiotics
- Antacids (especially magnesium-containing) and proton-pump inhibitors
- Chemotherapy agents
- Antihypertensives
- Bile acid sequestrants
- NSAIDs
- Prostaglandins
- Various anti-retroviral medications

Diarrhea from antibiotics should resolve within a few days of completion of the antibiotics. If the diarrhea persists, the patient should be tested for *C. difficile* infection (6).

What Fluid Intake Should Be Recommended?

Most patients who can drink fluids can avoid medical care for acute diarrhea. Fluids, electrolytes, and sugar need to be replaced, particularly in watery diarrhea. Patients without dehydration can easily be supplemented with water (or any non-caffeinated drink) and saltines or broth (3). In patients with more severe diarrhea and elderly patients, electrolytes should be replaced as well. Although Gatorade and other "sports drinks" do not contain the electrolyte levels that oral rehydration fluids, such as Pedialyte do (see Table 6-7) (6), they are readily available and far more palatable. Gatorade is isotonic, whereas Powerade and Allsport are hypertonic with more sugar and less sodium. Adding salt to Gatorade will make it hypertonic, which may increase the patient's diarrhea.

Oral rehydration formula has more electrolytes per liter than sports drinks and is ideal in cases of severe diarrhea. This is used in developing nations as a standard for fluid replacement. Commercial preparations such as Pedialyte and generic equivalents are available at most pharmacies. A solution can also be prepared at home with common household ingredients:

- 1 L water
- ½ tsp salt
- ½ tsp baking powder
- 4 tbsp sugar

Table 6-7　Comparison of Commercially Available Fluids

Component	Gatorade	Pedialyte
Sodium	23.5 mEq/L	45 mEq/L
Potassium	<1 mEq/L	20 mEq/L
Chloride	17 mEq/L	35 mEq/L
Citrate	Not available	30 mEq/L
Sugar	58 g sugar/L	25 g/L
Osmolality	280–360 mOsm/kg	. . .

Data from Reference 6.

As important as which fluid is how much. The short answer is the same amount as is lost in the diarrhea. Generally, patients should drink 1 to 2 liters in the first hours after the diarrhea begins, and then 1 to 2 liters daily until the diarrhea resolves. A good guide is the urine output; the patient should void dilute urine every 3 to 4 hours (8).

What Food Intake Should Be Advised?

Patients may believe that they either need to make themselves eat or that they should not eat at all. Fluids are clearly most important. The B.R.A.T. diet (bananas, rice, applesauce, and toast) is commonly recommended. The ACG recommends that patients with acute diarrhea eat a diet consisting of boiled starches, such as potatoes and rice, with salt added.

What Anti-Diarrheal Medications Should Be Recommended?

Patients with bothersome symptoms or dehydration should be guided in the use of anti-diarrheal medications (Table 6-8). Nausea should be treated as well, especially when it results in decreased oral fluid intake. All the following are available without a prescription and in generic forms.

Over the phone, bismuth subsalicylate (Pepto-Bismol and others) is a good first choice: it is effective, inexpensive, and fairly safe. It has been shown to reduce the number of stools by 50%, and has the added benefit of alleviating nausea and vomiting (3). Patients may already have this in the house. Another common resident of the home medicine cabinet is attapulgite (Kaopectate and others). This medication makes stools more formed. It contains no salicylates.

The opioids (loperamide and diphenoxylate/atropine) are generally not first choices over the telephone because: 1) they may prolong the duration of invasive diarrhea, 2) may increase the risk that *C. difficile* infection results in toxic megacolon, and 3) in the case of enterohemorrhagic *E. coli* (EHEC), may increase the risk of hemolytic-uremic syndrome (3). Despite these concerns, some patients will have these on hand, and may already be using them. For this reason, they are included in the following discussion.

Table 6-8　Comparison of Anti-Diarrheal Medications

Medication	Mechanism	Dose	Side Effects	Contraindications
Bismuth subsalicylate (Pepto-Bismol and others)	Anti-secretory (salicylate); anti-bacterial (bismuth); binds exotoxins and reduces vomiting	2 tabs or 1 tsp every 30–60 min up to 8 doses/day	• Dark stools (may mimic melena) • Darkened tongue	• Renal failure • Immunocompromised (risk of bismuth encephalopathy) • Aspirin/NSAID allergy • Influenza • Chicken pox
Attapulgite (Kaopectate and others)	Absorbs water, making stools more formed	2 tabs after every loose bowel movement, not to exceed 12 in a day		Not absorbed, very safe
Loperamide (Imodium, Imodium A-D, Kaopectate II, Pepto Diarrhea Control)	Opiate derivative; slowed intestinal transit time and peristalsis	2 mg capsule: 2 caps initially, followed by 1 cap after each loose stool up to 8 caps/day	• Nausea • Vomiting • Constipation • Abdominal distension • Xerostomia	• Bloody diarrhea • Hypersensitivity • Pseudomembranous colitis • EHEC
Diphenoxylate/ Atropine (Lomotil)	Opiate derivative; actions` same as loperamide; anticholinergic subtherapeutic dose included to discourage abuse	2.5 mg (diphenoxylate) tabs: 2 tabs 3–4 times/day	Similar to loperamide, plus: with MAO inhibitors: may potentiate hypertensive crisis; with alcohol: CNS depression; with antimuscarinics: may cause paralytic ileus	• Bloody diarrhea • EHEC • Pseudomembranous colitis • Hypersensitivity • Severe liver disease • Dehydration

Data from References 3 and 12.

Loperamide (Imodium and others) is generally effective with 80% reduction in stool number (3) and a more favorable side effect profile than the less expensive diphenoxylate/atropine (Lomotil and others). Again, these should be avoided if there are signs of an invasive pathogen.

H_2 blockers reduce nausea, so they should be considered as well when vomiting is a contributor to the patient's volume loss.

Is There Any Role for Antibiotics in the Home Treatment of Acute Diarrhea?

The use of empiric antibiotics in patients with acute diarrhea is controversial. Studies looking at the effect of antibiotics in unselected patients show minimal benefit (9). Certain authors recommend the empiric use of antibiotics in patients with blood in the stool because this is fairly specific for invasive diarrhea. Because of the controversy over the use of empiric antibiotics, it is not recommended to prescribe them to patients over the telephone, with the exception of patients with traveler's diarrhea (see box on page 126, Traveler's Diarrhea).

WHAT TO TELL THE PATIENT

High Risk: Patient Needs Emergency Evaluation

"It sounds to me like you are quite sick. I think you are very dehydrated or may have a serious infection that requires antibiotics." Or "I think you may have a very serious illness. The reason I think this is...." Tell the patient what warning signs are present and what complications you think the patient is at risk for.

"I recommend that you go to the emergency room as soon as possible." Recommend calling 911 if you think it is indicated, or instruct the patient to come to the office right away.

"You may need intravenous fluids and medication, which will make you feel a lot better and keep this from getting worse."

If the physician has facilities to provide intravenous fluids and obtain lab tests in the office, and the hospital is in close proximity, the physician may recommend immediate evaluation in the office.

Moderate Risk: Patient Needs Regular Telephone Follow-Up and May Need Urgent Evaluation

"It sounds like you're pretty sick with this, and you may need to come in to see me in the next day or two." Tell the patient what symptoms make you think this, and give a specific day and time to come in. "Right now I think there are some things you can do at home that will keep this from getting worse and help you feel better."

"Most people with your type of diarrhea do fine, but you'll need to be pretty careful to make sure you get enough fluids and salts. If you get dehydrated, you could get very sick. Severe dehydration can make you confused and can impair the function of your kidneys and other vital organs, so it's very important that we stay on top of this."

- *Choice of fluids:* Based on your assessment, recommend an appropriate fluid (water, sports drink, oral rehydration fluid) based on the degree or risk of dehydration. Explain why you are recommending a specific fluid.

- *Quantity of fluids:* "You're going to need to drink a lot more than you're probably used to. You need to keep up with what you're losing in diarrhea. I recommend at least 1 or 2 quarts a day, but you may need more. The best guide is your urine. You should drink enough so that you have to urinate every 3 to 4 hours, and it should come out clear."

- *Anti-diarrheal medication:* "I recommend you take something to stop the diarrhea." Providing a well-known brand name such as Pepto-Bismol allows quicker recognition by the patient, but generic products are less expensive and equally effective. Provide information on dosing and side effects (see Table 6-8).

- *Have the patient avoid being alone, if possible:* "It's a good idea for someone to stay with you as much as possible over the next 24 hours to help make sure you're getting enough fluids in and to help if you get sicker." Offer to speak with the companion.

- *When the patient should call back:* "If you start to feel worse, or feel dizzy, confused or very tired, or if you are unable to keep down the fluids, call me back right away. If you develop a fever, severe abdominal pain, or bloody diarrhea, call back right away."

- *Follow up in the first 24 hours:* "We should talk on the phone again in a few hours to make sure you're getting better. Do you think you can call me back at about...." Give a specific time because it may reduce the patient's reluctance to bother you again if the patient knows you are expecting a call. It also lets you know when to call the patient yourself if the patient hasn't called you yet.

Low Risk: Needs Reassurance but No Follow-Up

• *Probable benign nature of illness:* "It sounds like your diarrhea is probably from an infection, most likely a virus, and should resolve on its own. You don't seem to be dehydrated."

• *When the patient should call back:* "If you develop bloody diarrhea, severe pain, or fever, call back right away."

• *What patient should eat and drink:* "You should eat lightly and drink plenty of fluids, enough to make your urine clear; if it's not clear, you're not drinking enough fluid. Try to drink at least 2 quarts of fluid a day. If you want something for your symptoms, you may try a bismuth product such as Pepto-Bismol or another brand." Explain dosing and side effects (see Table 6-8).

Low Risk Acutely, Possible Underlying Pathology: Patient Needs Reassurance and Follow-Up Appointment Non-Urgently

Provide same recommendations as in Low Risk (see above), but add recommendation to follow-up. "While I don't think this episode is dangerous, the fact that you are having [small amount of bleeding/recurrent episodes] makes me think you should come in to be evaluated in person within the next week."

If there is blood in the stool, the patient should definitely avoid opioid anti-motility agents (loperamide/diphenoxylate).

WHAT TO DOCUMENT

Document the following:

• All symptoms, especially the presence or absence of warning signs
• Whether you conclude the patient is at low, moderate, or high risk of complications

PATIENT WHO CALLS TO REPORT DIARRHEA

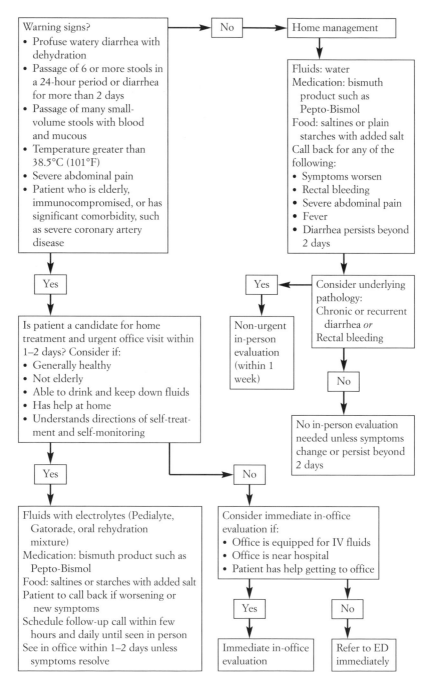

- What you told the patient, including the risk of complications, and if, when, and where they should be evaluated in person
- What you recommended in terms of fluids and medication
- When you told the patient to call back
- Your perception of how well the patient understood the instructions, including whether the patient repeated the instructions back to you and if he or she voiced agreement with the plan

WHEN TO CALL BACK

It is advisable to call the patient back any time the physician is uncertain about the patient's safety, either later that day or the next morning. The following guidelines should also be considered:

- *Low risk*—Physician usually does not need to call back.
- *Moderate risk*—Patient should be called back within a few hours to ensure appropriate fluid intake and that dehydration has not worsened.
- *High risk*—Physician should call back soon to ensure patient has gone to the emergency department (calling the emergency department later on would serve the same purpose).

REFERENCES

1. **Garthwright W, Archer D, Kvenberg J.** Estimates of incidence and cost of intestinal infectious diseases in the United States. Public Health Rep. 1988;103:107-15.
2. **Feldman R, Banatvala N.** The frequency of culturing stools from adults with diarrhoea in Great Britain. Epidemiol Infect. 1994;113:41-4.
3. **DuPont HL.** Guidelines on acute infectious diarrhea in adults. The Practice Parameters Committee of the American College of Gastroenterology. Am J Gastroenterol. 1997;92:1962.
4. **Johnson PC, Ericsson CD.** Acute diarrhea in developed countries: a rational for self-treatment. Am J Med. 1990;88:6S-9S.
5. **Koplan JP, Fineberg HV, Ferraro MJ, Rosenberg ML.** Value of stool cultures. Lancet. 1980;2:413-6.
6. **Aranda-Michel J, Gianella R.** Acute diarrhea: a practical review. Am J Med. 1999;106:670-6.
7. **McGee S, Abernethy WB 3rd, Simel DL.** Is this patient hypovolemic? JAMA. 1999;281:1022-9 .

8. **Bennett RG.** Acute gastroenteritis and associated conditions. In: Barker LR, Burton JR, Zieve PD, eds. Principles of Ambulatory Medicine, 5th ed. Philadelphia: Lippincott Williams & Wilkins; 1999:319-29.

9. **Wistrom J, Jertborn M, Ekwall E, et al.** Empiric treatment of acute diarrheal disease with norfloxacin: a randomized, placebo-controlled study. Swedish Study Group. Ann Intern Med. 1992;117:202-8.

10. **Steffen, R.** Epidemiological studies of travelers' diarrhea, severe gastrointestinal infections and cholera. Rev Infect Dis. 1986;8:S122.

11. **DuPont, HL, Ericsson, CD.** Prevention and treatment of travelers' diarrhea. N Engl J Med. 1993;16:616.

7

UPPER RESPIRATORY INFECTIONS AND RELATED ILLNESSES

Anna B. Reisman, MD

KEY POINTS

- Patients with upper respiratory infections (URIs) often call because they want antibiotics. Uncomplicated URIs do not require antibiotic treatment.
- Patients with warning signs of more serious complications of URIs, such as stridor, drooling, difficulty swallowing, severe dyspnea, or severe headache, should be referred for immediate emergency department (ED) evaluation.
- Patients with URI symptoms who may have pneumonia should be evaluated the same day either in the ED or in the office. Patients meeting any criteria for admission for pneumonia (altered mental status, severe vital sign abnormality), or with risk factors for increased morbidity and mortality in pneumonia should be evaluated in the ED. Patients with milder symptoms can generally be evaluated in the office.
- Non-emergency in-person evaluation should be considered for patients with URI symptoms consistent with possible bacterial sinusitis (maxillary toothache, colored nasal discharge, and poor response to nasal decongestants).
- Non-emergency in-person evaluation should be considered for patients with URI symptoms and symptoms consistent with acute otitis media (ear pain or discharge).

Continued

- Symptomatic treatments should focus on the predominant symptoms, and combination treatments should generally be avoided.
- There is no evidence for empirically treating patients with symptoms of influenza over the telephone with antiviral medications.

In the telephone evaluation, it is important to distinguish between patients with simple URIs that can be self-treated and those with complications that should be evaluated in person.

Most URIs are self-diagnosed, self-treated, and self-limited. The typical symptoms of the common cold (nasal obstruction, rhinorrhea, and sneezing, with or without a brief sore throat, often followed by the development of cough) are generally well known to both physicians and patients. Symptoms last usually 3-7 days (1), although 25% may last as long as 2 weeks (2). Common complications of URI and associated illnesses are listed in Table 7-1.

Physicians may wonder why patients call when they know they have a common cold. Some may want therapy prescribed, either because a prescription medicine may be less expensive than an over-the-counter treatment; some may seek a physician's advice rather than having to face an overwhelming selection of symptomatic medications at the drugstore; some may wonder whether a complication may have developed; others may want a definitive diagnosis; and yet others may hope to receive a prescription for antibiotics, which they believe will make them feel better more quickly (3). Many patients believe antibiotics are effective for the common cold (4).

Table 7-1   Complications of Upper Respiratory Infections

Common	*Less Common*
Acute sinusitis	Streptococcal pharyngitis
Acute otitis media	Pneumonia
Acute bronchitis	Bronchitis
Exacerbation of asthma or COPD	Epiglottitis

This chapter focuses on the following central questions:

- Should the patient be seen immediately? Does the patient have any warning signs of more serious complications?
- Could the patient have a bacterial infection requiring antibiotics?
- Could the patient have allergic rhinitis or influenza?
- How should a patient with an uncomplicated URI be treated at home?

BACKGROUND

Epidemiology

Viral upper respiratory infection is the most common acute illness in the United States. Adults, on average, have two to three URIs per year. It is the most common cause of absences from work and school and is associated with high costs of more than $3.5 billion per year (2,5).

URIs are spread by direct contact and aerosol inhalation. The most common etiologies are rhinoviruses, coronaviruses, and respiratory syncytial virus (RSV); influenza, parainfluenza, and adenoviruses may cause similar symptoms but are more typically associated with lower respiratory problems (2).

Utility of Early Diagnosis

Treating patients with uncomplicated URI over the telephone can have benefits for both patient and physician. The patient can avoid an office visit; the physician can avoid having more URIs transmitted to other patients and staff members in the office.

Early recognition of a complication of URI, such as acute sinusitis, pneumonia, or recurrent rheumatic fever, can prevent further clinical deterioration. Bacterial sinusitis develops in 0.5%-2.5% of adult patients after viral URI (6). Possible sequelae of acute sinusitis other than recurrence are rare and include osteomyelitis, mucocele, and invasion of the central nervous system. In adults with asthma, viral URIs may trigger up to 40% of acute asthma attacks (7).

Elderly patients and people with chronic illnesses are at risk for developing pneumonia after a URI. Up to 2%-9% of cases of pneumonia in the elderly follow an infection with RSV. RSV can also lead to exacerbations of CHF and other chronic conditions in the elderly (8).

Immunocompromised patients are also at higher risk of complications from URI. RSV causes a higher incidence of sinusitis and lower respiratory tract infection in these patients, with mortality of 11%-78% (9), and parainfluenza virus can cause life-threatening croup, bronchiolitis, and pneumonia (6).

APPROACH TO THE PATIENT WHO CALLS WITH UPPER RESPIRATORY INFECTION SYMPTOMS

Initially, the physician on the telephone should attempt to rule out the most dangerous sequelae of URI, as outlined below. The physician should then determine if the symptoms are suggestive of any complications of URI such as pneumonia, acute bronchitis, acute sinusitis, or acute otitis media, which might require in-person evaluation. If these are deemed unlikely, the physician should determine whether the symptoms are more likely to be associated with allergic rhinitis or influenza. For the patient with an uncomplicated URI, the physician can recommend self-care directed at particular symptoms.

SHOULD THE PATIENT BE SEEN IMMEDIATELY? DOES THE PATIENT HAVE ANY WARNING SIGNS OF MORE SERIOUS COMPLICATIONS?

Patients with any of the symptoms in Table 7-2 should generally be referred for ED evaluation. These symptoms are typical for more serious complications of URI such as epiglottitis, severe chronic obstructive pulmonary disease (COPD), or asthma exacerbation (see Chapter 16), severe pneumonia, and uncommon but potentially life-threatening complications of sinusitis-like meningitis and encephalitis.

Table 7-2　Warning Signs of Serious Complications of URIs

- Stridor
- Drooling
- Difficulty swallowing

- Moderate-to-severe dyspnea
- Severe headache

Table 7-3　Risk Factors for Increased Morbidity and Mortality in Pneumonia

- Hospitalized in past year
- Age > 65
- Coexisting illness (alcohol abuse, smoker, diabetes mellitus, immunosuppression, neoplastic disease, renal failure, heart failure, chronic lung disease, chronic liver disease, functional asplenia)
- Severe vital sign abnormalities
- Suspicion of aspiration
- Social considerations
- Patients with significant asthma
- Institutionalized elderly patient

COULD THE PATIENT HAVE A BACTERIAL SUPERINFECTION REQUIRING ANTIBIOTICS?

Could the Patient Have Pneumonia and Require Emergency Attention?

Patients who call with shortness of breath and other clinical features suggesting possible pneumonia (fever, cough, sputum production, dyspnea, fever, increased respiratory rate, pleurisy), and especially those with risk factors for increased morbidity and mortality in pneumonia, should generally be evaluated in the ED or as an emergency office visit (Table 7-3).

One group reviewed the difficulty inherent in diagnosing pneumonia by history alone (10). The authors looked at test characteristics of individual history items (discussed below) in the diagnosis of pneumonia, from four studies (11-14). No individual items were found to be significant in terms of their presence or absence helping to either rule in or rule out the diagnosis of pneumonia. The presence of certain findings such as fever, immunosuppression, and dementia may be helpful in indicating the need for further in-person evaluation but are not useful on their own because of the generally low prevalence of pneumonia in the study populations. Patients with symptoms suggestive of pneumonia with clinical features meeting criteria for admission for pneumonia (altered mental status, severe vital sign abnormality (pulse rate over 140, systolic blood pressure under 90, respiratory rate greater than 30/minute) should go to the ED (15).

Algorithms and prediction rules developed for helping determine which patients with suspected pneumonia should have chest radiographs have generally included elements of both the history and physical examination (10,11,13,14), which render them inapplicable to the telephone exam. One algorithm, which assigned points to symptoms and signs such as absence of rhinorrhea and sore throat, presence of night sweats, myalgias, sputum all day, respiratory rate greater than 25/minute, and temperature greater than 37.8° Celsius, may in some cases be useful on the telephone for patients who are able to give accurate temperature and respiratory rate readings (11) (see box, How Reliable Are Vital Signs Over the Telephone?). However, because pneumonia is uncommon in patients with URI symptoms, it is unlikely that such guidelines could definitively rule in pneumonia. Despite these problems, such guidelines can be useful in providing information for determining if a patient with suspected pneumonia may be at high risk for complications. Such high-risk patients, as discussed above, should be evaluated in the ED.

Physicians dealing with a patient with possible pneumonia on the telephone should use their clinical judgment in determining how rapidly and where a patient should be seen, whether immediately in the ER or in the office the same day, and should be especially conservative with patients with comorbid conditions. Generally, patients meeting any criteria for hospital admission and those with risk factors for increased morbidity and mortality in pneumonia should be evaluated in the ED; those with milder symptoms should still be evaluated in person within 24 hours.

HOW RELIABLE ARE VITAL SIGNS OVER THE TELEPHONE?

Some patients, family members, home care workers, or visiting nurses may call with vital signs measurements; the reliability of these signs, when collected over the phone by instructions to patients, has not been validated. Various studies have shown inaccuracies in how patients read thermometers and report subjective fevers (40,41). Similarly, accuracy of the measurement of respiratory rate has not specifically been studied over the telephone, to our knowledge. In some telephone cases, it might be possible to clearly hear an elevated respiratory rate; such information might be combined with other symptomatic information to help make a decision. In one study, the absence of any vital sign abnormalities was shown to decrease the predicted probability of pneumonia in ambulatory patients with respiratory illnesses (for a 5% prevalence, the predicted probability would be <1%) (10,12), but this would be difficult to assess over the phone.

Could the Patient Have Acute Bronchitis?

Acute bronchitis is usually manifested by productive cough and upper airway infection. Fever is unusual in acute bronchitis, and, when present, may be caused by influenza or pneumonia (16). If fever is present, immediate in-person evaluation is usually indicated for physical examination and possible chest x-ray to determine whether pneumonia could be present.

Patients without underlying chronic bronchitis with symptoms of acute bronchitis should be treated symptomatically and not with antibiotics because virtually all cases are viral (16,17,21). Although one recent review found that antibiotics had some benefits in acute bronchitis, there was a similar rate of adverse effects, and patients with other common cold symptoms for less than a week were unlikely to benefit from antibiotics (17). If there is any question whether the patient might benefit from antibiotics, in-person evaluation is warranted.

Patients with underlying chronic bronchitis, on the other hand, may benefit from antibiotics, although this is controversial. Such exacerbations usually present with an acute increase in usual COPD symptoms as listed in Table 7-4.

Table 7-4   Typical Symptoms of Acute Exacerbations of Chronic Bronchitis

---

• Increased volume or change in the character of sputum

• Increased frequency and severity of cough

• Increased dyspnea

• Variable constitutional symptoms

To our knowledge, no studies have been done on the usefulness of symptoms and signs suggestive of acute bronchitis over the telephone. If symptoms are consistent with an uncomplicated acute bronchitis in a patient without COPD without any warning signs of pneumonia, the patient can generally be treated symptomatically at home with a follow-up visit scheduled within a few days if symptoms persist. If symptoms and history suggest an exacerbation of chronic bronchitis, an office assessment is probably warranted within 24 hours, if possible, to rule out pneumonia and for possible antibiotic treatment. Some physicians may opt to treat a patient with known COPD with empiric antibiotics without an exam; however, there is no evidence to support such a decision.

Could the Patient Have Acute Bacterial Sinusitis?

The overall accuracy of individual symptoms in the diagnosis of sinusitis is poor (18). However, some findings might be helpful for a telephone examination in assessing whether sinusitis is more or less likely to be present. A study of 400 patients with suspected sinusitis found that absence of a preceding common cold was the most useful symptom in ruling out sinusitis, with a sensitivity of 85% (19). A study of 247 male patients with rhinorrhea or facial pain unrelated to trauma, or with suspected sinusitis, found that the most sensitive symptoms were colored nasal discharge (72%), cough (70%), and sneezing (70%) (20). Sore throat, itchy eyes, and constitutional symptoms were not useful in ruling out sinusitis. A study of 164 patients with suspected sinusitis found that six symptoms (preceding URI, nasal discharge, painful mastication, malaise, cough, and hyposmia) were significantly more common in patients with abnormal sinus x-rays, but no single finding was highly accurate (21). Toothache was the most useful

symptom in ruling in sinusitis in two studies in patients with suspected sinusitis (specificities 93% and 83%, respectively) (19-21).

An algorithm based on the accuracy of combinations of symptoms and
signs can be very useful to the physician trying to rule in or rule out acute
sinusitis in person, but is less helpful for the telephone physician (18,20).
Three symptoms (maxillary toothache, poor response to nasal decongestants, and history of colored nasal discharge) and two signs (purulent nasal
discharge and abnormal transillumination) were found to be the best predictors of acute sinusitis. When no findings were present, the likelihood
ratio (LR) was 0.1, effectively ruling out the diagnosis; when four or more
findings were present, the LR was 6.4. On the telephone, however, only
the three symptoms would be applicable, and the algorithm was not validated for these elements alone.

The absence of certain symptoms (preceding common cold, colored
nasal discharge, cough, sneezing) can help the physician rule out sinusitis,
although none of these is highly sensitive or accurate. Non-emergency in-
person evaluation, assuming the absence of any warning signs, should be
considered for patients with URI symptoms consistent with possible bacterial sinusitis (maxillary toothache, colored nasal discharge, and poor response to nasal decongestants).

Could the Patient Have Acute Otitis Media?

In adults, the most common symptom of otitis media is ear pain.
Patients with ear pain associated with a URI should be seen in the office
within 24 hours for an exam and determination of whether antibiotics
are indicated. Otitis media can be reliably diagnosed only in person.
The few studies looking at telephone diagnosis of acute otitis media are
from the pediatric literature and have shown it to be inaccurate. In a
study from the pediatric literature that looked at the accuracy of telephone diagnosis of otitis media, diagnostic error was found in 20% of
cases (22,23).

Could the Patient Have Streptococcal Pharyngitis?

Patients with sore throat complicated by respiratory compromise should
be evaluated immediately; for other patients, clinical symptoms and signs

are not specific enough to reliably distinguish streptococcal from non-streptococcal pharyngitis, and diagnosis depends on laboratory testing. Patients with symptoms suggestive of streptococcal pharyngitis should be evaluated in person on a non-emergency basis with a throat culture or rapid antigen test. Chapter 10 provides information on the evaluation of sore throat over the telephone.

COULD THE PATIENT HAVE ALLERGIC RHINITIS OR INFLUENZA?

Patients with allergic rhinitis or influenza may present similarly to patients with URIs. Table 7-5 compares common presentations of these three diagnoses. Patients with allergic rhinitis often have a clear nasal discharge with sneezing, sniffing, nasal itchiness, and itchy, red, watery eyes.

Patients with influenza usually present with respiratory tract symptoms in combination with the abrupt onset of fever, headache, myalgias, and malaise during the flu season. Influenza is very unlikely to occur during non-winter months. If the patient has symptoms during

Table 7-5 Comparison of Typical Symptoms of Upper Respiratory Infections, Influenza, and Allergic Rhinitis

Upper Respiratory Infection	Influenza	Allergic Rhinitis
• Gradual onset of symptoms	• Abrupt onset of symptoms	• Clear nasal discharge
• Possible mild fever	• High fever lasting 3–4 days	• Sniffing
• Nasal congestion		• Nasal itchiness
• Rhinorrhea	• Chills	• Itchy, red, watery eyes
• Mild-to-moderate body aches	• Headache	• Sneezing
	• Myalgia	• Seasonal
• Mild fatigue	• Malaise	• Previous episodes of similar symptoms
• Sneezing	• Dry cough	
• ± Brief sore throat	• Severe fatigue, weakness, exhaustion	
• Cough after waning of nasal symptoms	• Seasonal (winter months)	

seasons when influenza does not normally occur, the patient should be seen in person for laboratory diagnosis for confirmation. With uncomplicated influenza, symptoms usually improve over 2-5 days. Patients who have flu-like symptoms but report having had a flu shot are still at risk for flu.

What are Typical Complications of Influenza?

Pneumonia, either primary influenza pneumonia or secondary bacterial pneumonia, is the most common complication of influenza. It occurs primarily in high-risk groups (patients with chronic cardiovascular or pulmonary illnesses, diabetes mellitus, renal disease, hemoglobinopathy, or immunosuppression; patients who live in nursing homes or who are over 65 are also at higher risk) (42). Primary influenza pneumonia should be suspected when typical flu symptoms persist and increase in severity, or when dyspnea becomes apparent. Secondary bacterial pneumonia should be suspected when the symptoms initially improve, followed by an exacerbation of fever and respiratory symptoms. When these are suspected, patients should be seen urgently for evaluation.

Should the Patient with Likely Uncomplicated Influenza Be Treated Empirically at Home with Antiviral Medication?

Most patients with uncomplicated flu can be treated symptomatically. There is no evidence for empirically treating patients with probable uncomplicated flu over the telephone with antiviral medications such as the neuraminidase inhibitors zanamivir and oseltamivir. For those in whom treatment is considered, it should be started within 2 days of the onset of symptoms; some authors recommend continuing treatment only in patients with laboratory-proven influenza (43). Several studies showed that these treatments reduced the duration of symptoms by about 1 day, and were more effective in patients with more severe symptoms (44,45). These studies were done primarily on healthy adults; the role of these medications in high-risk groups has yet to be defined. These agents may be indicated in patients who have not been vaccinated, especially those with chronic diseases, and in those who develop influenza despite vaccination (46).

WHAT KINDS OF TREATMENT SHOULD BE RECOMMENDED FOR THE PATIENT WITH A SIMPLE UPPER RESPIRATORY INFECTION?

There are many symptomatic treatments for URI available over the counter as well as via prescription. Generally, it is recommended to focus the treatment on the predominant symptoms, rather than telling the patient to use a combination treatment, which may have more side effects.

Nasal Congestion, Rhinorrhea, and Sneezing

Ipratropium Bromide (Intranasal)

This intranasal anticholinergic medication has been shown to lessen symptoms of rhinorrhea and sneezing (24). Side effects include nasal dryness, higher rates of blood-tinged mucus, and headache.

Cromolyn Sodium (Intranasal)

In a study in which patients with cold symptoms for less than 24 hours were treated with cromolyn sodium, symptoms resolved faster and were less severe in the last three days of treatment than in those who received placebo (25). Side effects were mild with no significant difference between the two groups; however, at least one author has pointed out that the drug has not yet been studied in larger populations (26).

Antihistamines

The efficacy of antihistamines in the common cold is controversial. Several first-generation antihistamines have been shown to decrease rhinorrhea, secretions, and sneezing (26-28); however, some of these studies have been found unreliable in terms of their methodology (29). Although the common side effects of antihistamines (sedation, dry nose and mouth) may limit their usefulness, some patients find the sedating effect useful if taken before bed.

Oral Decongestants

Medications such as pseudoephedrine may be effective in lessening the nasal congestion of the common cold. Such medications may be contraindicated in some patients with poorly controlled hypertension or coronary artery disease.

Topical Decongestants

Medications such as naphazoline and oxymetazoline may alleviate nasal congestion. These medications should not be used for more than 5 days because of the possibility of causing rebound congestion and rhinitis medicamentosa. Because of their sympathomimetic action, they should be used with caution in patients with hypertension.

Fevers, Aches, and Pains

Analgesics

Analgesics such as aspirin, acetaminophen, and ibuprofen can help with symptoms of fever, headache, myalgia, and sore throat. However, they have been shown to cause modest reductions in immune function and to cause an increase in nasal symptoms and signs (30). Another study found that naproxen was beneficial for headache, malaise, myalgia, and cough without affecting the antibody response (31).

Cough

Antitussive Therapy

Because the cough from the common cold is usually from post-nasal drip, decongestants may be more helpful than cough suppressants.

Symptom Duration

Zinc Lozenges

The evidence for zinc lozenges as a treatment for URI is inconclusive (5,32). Although zinc lozenges may slightly decrease the duration of the common cold if therapy is begun within the first 24 hours (33), a meta-analysis of six trials found no consistent evidence that they reduced the duration of the common cold (6). Zinc lozenges are difficult to take: they must be taken every 2 hours while awake, and side effects of nausea and bad taste may occur as well.

Vitamin C

The use of vitamin C in the treatment of the common cold remains controversial. One recent review of the literature found that relatively high doses of vitamin C had some benefit in reducing the duration of cold symptoms (34). An earlier review found an average 23% reduction in symptom duration (35). The optimal dose for treatment of the common cold remains unclear, however, because different doses of vitamin C were used in the various studies.

Other Treatments

Echinacea

The botanical echinacea is often used to prevent or treat URIs, although the current literature is conflicting (5). One review of placebo-controlled prevention trials reported that the majority of studies reported positive results, but there was not enough evidence to recommend a specific product (36). Another review questioned the validity of positive results because of methodologic problems in several studies (37).

Antibiotics

The majority of patients with uncomplicated URIs will not benefit from antibiotics. There is no indication to treat a URI over the phone with antibiotics. If a complication of a URI is suspected where antibiotic treatment may be appropriate, the patient should be evaluated in person.

Other Remedies

There is no evidence that bed rest, steam inhalation, salt-water gargles, or saline nasal sprays improve URI symptoms, although none of these is likely to have any significant side effects. Voice rest is recommended for hoarseness.

HOW CAN THE PHYSICIAN TEACH A PATIENT ABOUT SELF-CARE OF UPPER RESPIRATORY INFECTIONS AND HOW TO PREVENT THEM IN THE FUTURE?

During the winter months, the volume of calls regarding URIs frequently increases the workload of the physician and office staff. Several studies have looked at the efficacy of giving patients information on self-treatment of URIs, either written materials via the mail or face-to-face counseling. In one, a mailed self-care pamphlet (mailed without individual counseling) showed no significant effect on reducing telephone calls or office visits (38). Another study found that the combination of educational self-care materials plus individual counseling reduced unnecessary visits for URIs by 44% (39).

It may therefore be helpful to teach patients (and family members) about preventing colds by handwashing, and it may also be helpful to follow up the phone call with a self-care pamphlet on URIs. Patients should be taught about the symptoms of viral URI, how to recognize complications of URIs, and self-care of URIs.

WHAT TO TELL THE PATIENT

• *Warning signs present:* "I am concerned that you could have [state possible serious complication]. You need to go to the emergency department immediately."

• *Uncomplicated URI:* "It sounds like you have a [bad] cold. Colds are caused by viruses, and there is no treatment that can cure them. Antibiotics will not treat a cold. I can recommend some things that will help with your symptoms. [Give instructions based on patient's predominant symptoms.] I also recommend that you wash your hands frequently and that you advise the people you live with to do the same. If you develop any warning signs of a more serious infection [such as drooling, difficulty swallowing, breathlessness, severe headache], you should call back right away."

• *Patients who insist on antibiotics:* "The common cold is caused by a virus and not bacteria. Unfortunately, antibiotics don't stop viruses, so antibiotics are not going to help; in fact, they may cause more problems from side effects or lead to problems in the future due to resistance. You are

PATIENT WHO CALLS WITH SYMPTOMS OF UPPER RESPIRATORY INFECTION

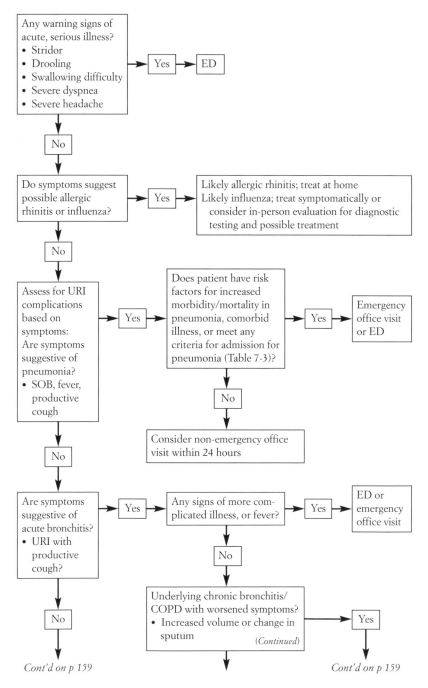

Cont'd on p 159

Cont'd on p 159

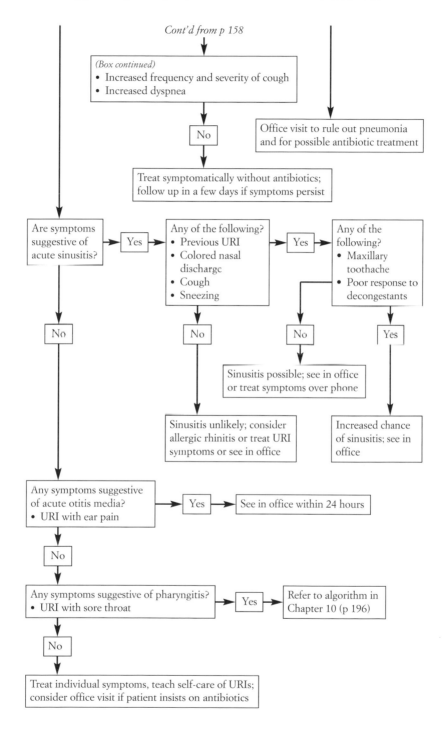

Cont'd from p 158

(Box continued)
• Increased frequency and severity of cough
• Increased dyspnea

No

Office visit to rule out pneumonia and for possible antibiotic treatment

Treat symptomatically without antibiotics; follow up in a few days if symptoms persist

Are symptoms suggestive of acute sinusitis?

Yes

Any of the following?
• Previous URI
• Colored nasal discharge
• Cough
• Sneezing

Yes

Any of the following?
• Maxillary toothache
• Poor response to decongestants

No

No

No

Yes

Sinusitis possible; see in office or treat symptoms over phone

Sinusitis unlikely; consider allergic rhinitis or treat URI symptoms or see in office

Increased chance of sinusitis; see in office

Any symptoms suggestive of acute otitis media?
• URI with ear pain

Yes

See in office within 24 hours

No

Any symptoms suggestive of pharyngitis?
• URI with sore throat

Yes

Refer to algorithm in Chapter 10 (p 196)

No

Treat individual symptoms, teach self-care of URIs; consider office visit if patient insists on antibiotics welcome to come in for an examination in person if you prefer." The physician can try to explain the lack of need for antibiotics, but some patients may be more satisfied with an office visit where they receive a physical exam and reassurance that appropriate signs for illness requiring antibiotics were sought (3).

• *Possible sinusitis, otitis media, streptococcal pharyngitis:* "It sounds like your cold may have developed into an infection of your [sinuses/ear infection/throat]. I would like for you to come in to the office today or tomorrow so I can perform a physical exam and order appropriate tests. This will help me decide whether antibiotics will be likely to help."

• *Possible pneumonia:* "I'm concerned that you may have developed pneumonia. The only way to know for certain is to have a physical examination and a chest x-ray, and it is very important that we find out whether this is the case. I would like for you to [go to the emergency department/come to the office] to be evaluated."

WHAT TO DOCUMENT

Document the following:

- Specific symptoms, relevant past history, and current medications
- Pertinent positives and negatives considered as possible serious complications of URIs
- Any risk factors for increased morbidity or mortality
- Working diagnosis
- Medications prescribed, if any; any recommendations for specific over-the-counter and non-pharmacologic treatments
- Clear follow-up plan
- Patient informed to call back with any significant change in symptoms

REFERENCES

1. **Kirkpatrick G.** The common cold. Prim Care. 1996;23:657-75.
2. **Turner R.** Epidemiology, pathogenesis, and treatment of the common cold. Ann Allergy Asthma Immunol. 1997;78:531-9.
3. **Olson L.** Is prescribing antibiotics by phone for respiratory infections acceptable? Postgrad Med. 1999;105:46-51.

4. **Wilson A, Crane L, Barrett P, Gonzales R.** Public beliefs and use of antibiotics for acute respiratory illness. J Gen Intern Med. 1999;14:658-62.

5. **Sexton D, Friedman N.** The common cold. UpToDate CD-ROM; 2000.

6. **Jackson J, Peterson C, Lesho E.** A meta-analysis of zinc salts lozenges and the common cold. Arch Intern Med. 1997;157:2373-6.

7. **Teichtahl H, Buckmaster N, Pertnikovs E.** The incidence of respiratory tract infection in adults requiring hospitalization for asthma. Chest. 1997;112:591-6.

8. **Han L, Alexander J, Anderson L.** Respiratory syncytial virus pneumonia among the elderly: an assessment of disease burden. J Infect Dis. 1999;179:25-30.

9. **Wendt C, Hertz M.** Respiratory syncytial virus infections in the immuno-compromised host. Semin Respir Infect. 1995;10:224-31.

10. **Metlay J, Kapoor W, Fine M.** Does this patient have community-acquired pneumonia? JAMA. 1997;278:1440-5.

11. **Diehr P, Wood R, Bushyhead J, et al.** Prediction of pneumonia in outpatients with acute cough—a statistical approach. J Chronic Diseases. 1984;37:215-25.

12. **Gennis P, Gallagher J, Falvo C, et al.** Clinical criteria for the detection of pneumonia in adults: guidelines for ordering chest roentgenograms in the emergency department. J Emerg Med. 1989;7:263-8.

13. **Singal B, Hedges J, Radack K.** Decision rules and clinical prediction of pneumonia: evaluation of low-yield criteria. Ann Emerg Med. 1989;18:13-20.

14. **Heckerling P, Tape T, Wigton R, et al.** Clinical prediction rule for pulmonary infiltrates. Ann Intern Med. 1990;113:664-70.

15. **Gertman P, Restuccia J.** The Appropriateness Evaluation Protocol: a technique for assessing unnecessary days of hospital care. Medical Care. 1981;19:855-71.

16. **Bartlett J.** Acute bronchitis. UpToDate CD-ROM; 2000.

17. **Smucny J, Fahey T, Becker L, et al.** Antibiotics for acute bronchitis. The Cochrane Library. 2000;4.

18. **Williams J, Simel D.** Does this patient have sinusitis? JAMA. 1993;270:1242-6.

19. **van Duijn N, Brouwer H, Lamberts H.** Use of symptoms and signs to diagnose maxillary sinusitis in general practice: comparison with ultrasonography. BMJ. 1992;305:684-7.

20. **Williams J, Simel D, Roberts L, Samsa G.** Clinical evaluation for sinusitis: making the diagnosis by history and physical examination. Ann Intern Med. 1992;117: 705-10.

21. **Axelsson A, Runze U.** Symptoms and signs of acute maxillary sinusitis. ORL J Otorhinolaryngol Relat Spec. 1976;38:298-308.

22. **Nickerson H, Biechler L, Witte F.** How dependable is diagnosis and management of earache by telephone? Clin Pediatr. 1975;14:920-3.

23. **Pichichero M.** Changing the treatment paradigm for acute otitis media in children. JAMA. 1998;279:1748-50.

24. **Hayden F, Diamond L, Wood P, et al.** Effectiveness and safety of intranasal ipratropium bromide in common colds: a randomized, double-blind, placebo-controlled trial. Ann Intern Med. 1996;125:89-97.

25. **Aberg N, Aberg B, Alestig K.** The effect of inhaled and intranasal sodium cromoglycate on symptoms of upper respiratory tract infections. Clin Ext Allergy. 1996;26:1045-50.

26. **Mossad S.** Fortnightly review: treatment of the common cold. BMJ. 1998;317:33-6.

27. **Halberstam M.** Medicine by telephone—is it brave, foolhardy, or just inescapable? Mo Med. 1977:11-15.
28. **Turner R, Sperber S, Sorrentino J.** Effectiveness of clemastine fumarate for treatment of rhinorrhea and sneezing associated with the common cold. Clin Infect Dis. 1997:824.
29. **Luks D, Anderson M.** Antihistamines and the common cold: a review and critique of the literature. J Gen Intern Med. 1996;11:240.
30. **Graham N, Burrell C, Douglas R, et al.** Adverse effects of aspirin, acetaminophen and ibuprofen on immune function, viral shedding and clinical status in rhinovirus-infected volunteers. J Infect Dis. 1990;162:1277-82.
31. **Sperber S, Hendley O, Hayden F, et al.** Effects of naproxen on experimental rhinovirus colds: a randomized, double-blind, controlled trial. Ann Intern Med. 1992;117:37-44.
32. **Marshall I.** Zinc for the common cold. Cochrane Database Syst Rev. 2000;2: CD001364.
33. **Mossad S, Macknin M, Medendorp S, Mason P.** Zinc gluconate lozenges for treating the common cold. Ann Intern Med. 1996;125:81-8.
34. **Douglas R, Chalker E, Treacy B.** Vitamin C for preventing and treating the common cold. Cochrane Database Syst Rev. 2000;2:CD000980.
35. **Hemila H, Herman Z.** Vitamin C and the common cold: a retrospective analysis of Chalmers' review. J Am Coll Nutr. 1995;14:116-23.
36. **Melchart D, Linde K, Fischer P, Kaesmayr J.** Echinacea for preventing and treating the common cold. Cochrane Database Syst Rev. 2000;2:CD000530.
37. **Giles J, Palat CR, Chien S, et al.** Evaluation of echinacea for treatment of the common cold. Pharmacotherapy. 2000;20:690-7.
38. **Stergachis A, Newmann W, Williams K, Schnell M.** The effect of a self-care minimal intervention for colds and flu on the use of medical services. J Gen Intern Med. 1990;5:23-8.
39. **Roberts C, Imrey P, Turner J, et al.** Reducing physician visits through consumer education. JAMA. 1983;250:1986-9.
40. **Buckley R, Conine M.** Reliability of subjective fever in triage of adult patients. Ann Emerg Med. 1996;27:693-5.
41. **Fletcher J, Creten D.** Perceptions of fever among adults in a family practice setting. J Fam Pract. 1986;22:427-30.
42. **Dolin R.** Clinical manifestations and diagnosis of influenza. UpToDate. 1999;8:1.
43. **Long J, Mossad S, Goldman M.** Antiviral agents for treating influenza. Cleve Clin J Med. 2000;67:92-5.
44. **Monto A, Fleming D, Henry D, et al.** Efficacy and safety of the neuraminidase inhibitor zanamivir in the treatment of influenza A and B virus infections. J Infect Dis. 1999;180:254-61.
45. **Treanor J, Hayden F, Vrooman P, et al.** Efficacy and safety of the oral neuraminidase inhibitor oseltamivir in treating acute influenza: a randomized controlled trial. JAMA. 2000;283:1016-24.
46. **Nichol K.** Commentary on Monto AS et al. ACP J Club. 2000;132:92-3.

8

DYSURIA IN WOMEN

Sally G. Haskell, MD

KEY POINTS

- 60%-70% of women with dysuria will have an uncomplicated urinary tract infection (UTI).
- Women with presumed uncomplicated cystitis should usually be referred for urinalysis or urine dipstick to confirm the diagnosis, although low-risk patients without diabetes, pregnancy, immunosuppression, or suspicion of complicated UTI whose symptoms suggest uncomplicated cystitis can sometimes be treated empirically.
- Patients with dysuria associated with fever, nausea, vomiting, and flank pain should be suspected of having pyelonephritis and should be referred for immediate evaluation and treatment.
- Patients with possible sexually transmitted diseases (STDs) causing urethritis should be referred for pelvic exam and cultures within 1-2 days.
- Patients with possible subacute pyelonephritis should be referred for urinalysis and culture before being treated.
- Most women with previously culture-documented UTI who telephone with recurrent symptoms can be treated empirically.

Dysuria is a frequent complaint among women calling to seek urgent treatment in internal medicine offices. Though most cases of dysuria are caused by uncomplicated cystitis, dysuria may also be one of the presenting signs in more serious infection, including pyelonephritis and pelvic inflammatory disease (PID).

Complications of untreated infection can include short-term morbidity and loss of time from work, recurrent infection, serious illness, hospitalization, bacteremia, and death. Timely and appropriate diagnosis and treatment is important in minimizing morbidity and mortality associated with UTIs. This chapter will address how the physician on the telephone can separate women with cystitis or vaginitis from those at risk of serious infection, and under what circumstances antibiotics should be initiated without an office visit. This chapter focuses on these two central questions:

1. Does the patient require urgent or emergency evaluation?
2. When can treatment be recommended over the telephone?

This chapter does not address the telephone management of dysuria in men.

BACKGROUND

Epidemiology

Two-thirds of women with dysuria will have lower UTI (1) accounting for more than 7 million office visits annually in the United States and affecting half of women at least once during their lifetime (2,3). A minority of patients with dysuria will have urethral infection with gonorrhea or chlamydia, or urethritis or vaginitis caused by trichomoniasis, candidiasis, or herpes simplex (1). Acute pyelonephritis occurs in 250,000 women in the United States each year (4).

Utility of Early Diagnosis

One of the main tasks of the physician who encounters a patient with dysuria on the telephone is to differentiate between uncomplicated cystitis or vaginitis and more serious infections, such as acute pyelonephritis and PID, which require immediate emergency treatment. Upper UTI is the most frequent source of community-acquired bacteremia (5), and severe UTI complicated by bacteremia has a mortality rate of 10%-20% (6). Tubal damage and scarring from PID can result in important long-term complications such as an increased risk for chronic pelvic pain (18%), ectopic pregnancy (six times more frequent), infertility (8% after one episode, 19.5% after two episodes, and 40% after three or more episodes), and subsequent episodes of PID (7,8).

Early diagnosis and treatment of uncomplicated UTI can reduce patient discomfort and can reduce cost and inconvenience.

DOES THE PATIENT REQUIRE EMERGENCY EVALUATION?

Because there are few data that specifically addresses the evaluation of dysuria over the telephone, the probability of each condition should be assessed by clinical symptoms and risk factors.

Patients whose symptoms suggest acute pyelonephritis, urosepsis, or PID should be evaluated as an emergency in person.

Could the Patient Have Acute Pyelonephritis?

Patients with acute pyelonephritis usually have classic signs and symptoms including dysuria, frequency and urgency in association with fever, flank pain, nausea, and vomiting. These symptoms have usually developed rapidly over hours to days. Fever associated with dysuria, frequency, and urgency may be one of the most sensitive indicators of upper UTI (9).

Could the Patient Have Pelvic Inflammatory Disease?

Patients with PID may call to report with symptoms of vaginal discharge, abdominal pain, fever, nausea, vomiting, dysuria, and dyspareunia. The most common presenting symptom is usually bilateral lower abdominal pain (10).

Could the Patient Have Urosepsis or Bacteremia?

Patients with dysuria who have diabetes and fever may have a higher likelihood of urosepsis (11). Symptoms and signs suggesting possible bacteremia include fever or hypothermia, chills, rigors, changes in mental status, and hyperventilation (12).

DOES THE PATIENT REQUIRE NON-EMERGENCY EVALUATION?

After excluding the most serious conditions, the physician should try to determine the cause of dysuria in order to decide if a "non-emergency" in-person evaluation is indicated. Patients with dysuria likely secondary to urethritis should generally be seen within 1-2 days, whereas patients with probable vaginitis can be seen, in most cases, for a routine office visit.

Could the Patient Have Urethritis?

Factors that should increase the suspicion of infectious urethritis from gonorrhea or chlamydia rather than a UTI include patients with a history of an STD, patients with new sex partners in the past few weeks, a sex partner with urethral symptoms, or symptoms that have begun gradually over several weeks (1). Women at risk for urethritis should be seen in person within 1-2 days for evaluation and culture because of the possibility of serious upper genital tract infection and long-term consequences, as discussed above. Although there are no specific data regarding the magnitude of increased risk with increasing delay in treatment of STDs, the serious long-term consequences of untreated infection suggest that treatment within a short time interval is warranted.

Could the Patient Have Vaginitis?

Women with vaginitis from bacterial vaginosis, *Trichomonas*, or *Candida* may present with symptoms of dysuria rather than vaginal discharge or irritation, but vaginal symptoms are almost always present in women with vaginitis and can usually be elicited with directed questioning (13). If a patient is felt to have dysuria caused by vaginitis and risk factors or symptoms for PID or if STDs have been excluded, she can be referred for a routine office visit for diagnosis and treatment. See Chapter 9 for more details on the telephone evaluation of vaginitis.

DOES THE PATIENT REQUIRE A URINALYSIS AND URINE CULTURE BEFORE TREATMENT IS INITIATED?

Could the Patient Have Subclinical Pyelonephritis?

Some patients with UTI without classic symptoms of pyelonephritis may still be at risk for subclinical pyelonephritis. These patients appear clinically to have lower UTI but have upper tract infection (demonstrable by ureteral catheterization, bladder washout techniques, and antibody-coated bacteria assays) (14). Symptoms are often mild and may smolder for long periods of time.

The importance of identifying these patients over the telephone is that they require urinalysis and culture for diagnosis and may need prolonged therapy to eradicate infection. Risk factors for this syndrome include underlying urinary tract abnormality, diabetes, immunocompromised condition, history of childhood UTIs, symptoms of cystitis for 7-10 days, acute pyelonephritis within previous year, relapse of infection (with the same organism) and elderly and institutionalized women (12).

These patients should be sent for a urinalysis and culture before initiating treatment, usually within 24 hours of their phone call.

Could the Patient Have a Complicated Urinary Tract Infection?

Complicated UTIs occur in patients with functionally, anatomically, or metabolically abnormal urinary tracts or in patients suspected of having UTI caused by resistant pathogens. Any patient suspected to have a complicated UTI should have microscopic urinalysis, culture, and sensitivity performed to identify pathogens and to direct treatment appropriately, within 24 hours.

WHEN CAN TREATMENT BE RECOMMENDED OVER THE PHONE?

After excluding the more serious conditions, the physician on the telephone should determine whether the patient could be treated empirically without an office or emergency room visit. For women who call to report dysuria, empiric treatment by telephone is appropriate only in the case of acute uncomplicated UTIs and in recurrences of such infections.

Could the Patient Have an Acute Uncomplicated Urinary Tract Infection?

Acute uncomplicated UTIs are manifested by dysuria, usually in combination with frequency, urgency, suprapubic pain, and/or hematuria. They occur most frequently in sexually active young women. The following factors increase the risk of infection (15,16):

- Recent sexual intercourse
- Use of a diaphragm and spermicide (and possibly spermicide alone)
- Delayed post-coital micturition
- History of recent UTI

In general, patients with uncomplicated acute UTIs do not have symptoms of vaginal discharge or irritation and do not have a history of risk factors for chlamydia or other causes of infectious urethritis (17).

Although most patients with probable uncomplicated UTI can undergo an abbreviated laboratory work-up with microscopic urinalysis or dipstick leukocyte esterase testing followed by empiric therapy (18), recent studies have shown that empiric treatment without urinalysis or urine culture can be effective as well in selected patients (19,20). Because the spectrum of infectious organisms (80% *E. coli*, 5%-15% *S. saprophyticus*, and occasional *Klebsiella* and *Proteus*) (21) that cause acute uncomplicated cystitis in young women is narrow, most organisms will be covered by one of several antibiotics. Empiric treatment, without laboratory documentation of UTI, may therefore be appealing to both physician and patient. It is also the most cost-effective option (19).

Table 8-1   Reasons for Exclusion from Telephone Management of Presumed Uncomplicated Urinary Tract Infection

- Diabetes mellitus

- Pregnancy

- Immunosuppression

- Suspicion of a complicated UTI (e.g., catheter-related)

- Suspicion of a sexually transmitted disease (e.g., presence of vaginal discharge)

Modified from Saint S, Scholes D, Fihn S. The effectiveness of a clinical practice guideline for the management of presumed uncomplicated urinary tract infection in women. Am J Med. 1999;106:637–41.

One recent study used a simple guideline to guide telephone management of women with dysuria (20). Women who telephoned or presented with dysuria were screened for the presence of any complicating factors (Table 8-1). If they had none, they were triaged to a primary care nurse and offered the choice of an office visit or telephone management. Forty percent were managed over the telephone. There was no significant increase in potential adverse effects, including revisits for cystitis, STD, and pyelonephritis within 60 days of the initial diagnosis. A telephone questionnaire of 100 randomly selected patients who were managed by the guideline showed that 95% of the patients were satisfied with the care they received, and 85% stated that for future UTIs they would rather have telephone treatment than an office visit.

WHAT EMPIRIC TREATMENT SHOULD BE RECOMMENDED FOR PATIENTS WHO ARE NOT PREGNANT?

A 3-day course of trimethoprim-sulfamethoxazole (160/800 mg every 12 hours) is recommended as first-line empiric treatment because it is effective in most cases and inexpensive. First-line treatment for patients who are allergic, or in areas with resistance to trimethoprim-sulfamethoxazole, should be a fluoroquinolone (for example, ciprofloxacin 100-250 mg every 12 hours or levofloxacin 250 mg once a day). A 7-day course of nitrofurantoin (nitrofurantoin monohydrate macrocrystals 100 mg every 12 hours) or a 3-day course of an oral broad-spectrum cephalosporin (for example, cefixime

400 mg once a day, or cefpodoxime proxetil 100 mg every 12 hours) can also be used. Patients with significant dysuria may also benefit from phenazopyridine (200 mg three times a day for 1-2 days as needed) for bladder analgesia; they should be told that the drug may turn their urine dark orange.

HOW SHOULD WOMEN WITH RECURRENT INFECTIONS BE MANAGED?

As many as 20% of women with an initial episode of cystitis experience recurrent infections (2). *Recurrent infection* is defined as reinfection with an exogenous organism, usually months apart, whereas *relapsing infection* indicates a persistent focus of infection and most often occurs within days to weeks after discontinuing treatment. Recurrent uncomplicated cystitis is rarely associated with anatomical or functional abnormalities of the urinary tract, and studies of excretory urography and cystoscopy have demonstrated anatomic abnormalities in less than 5% of women with recurrent UTI (23). Thus, routine evaluation of the urinary tract in such patients is not recommended. Recurrent cystitis should be documented by culture at least once (to distinguish it from relapsing infection) and subsequently can be reliably diagnosed over the telephone without performing a urinalysis or culture (24).

Women with previously documented recurrent UTIs who telephone with the same UTI symptoms can be treated empirically without urinalysis or culture (24). This approach should be limited to those women with a documented history of recurrent (not relapsing) infection, who have unequivocal UTI symptoms and no suspicion of complicated UTI. Patient-initiated therapy initiated when symptoms arise can provide a convenient, safe, inexpensive, and effective management strategy (25).

WHAT TO TELL THE PATIENT

High Risk
• *Possible acute pyelonephritis or urosepsis:* The patient should be referred directly to an emergency room. Ambulance transport may not be necessary unless the physician makes the assessment that the patient may have a life-threatening infection.

"You may have a serious infection, which can be life-threatening. You should go immediately to an emergency room. In the emergency room, the physicians will examine you, take blood and urine cultures and start appropriate treatment, which will prevent more severe infection and may save your life."

• *Possible subclinical pyelonephritis or STD:* The patient should be advised to seek treatment soon (within 24-48 hours) from her provider or in an emergency facility.

"You may have a serious infection that should be treated to prevent development of an even more serious infection or long-term complications. You should go to your provider or an urgent care facility within the day if possible."

• *What to do after you hang up:* Call back later. If the patient seems ambivalent, a call back may help convey your sense of urgency. If you have advised the patient to report to the emergency department, you should contact the emergency department. The primary care provider should also be notified.

Low Risk—If the patient's symptoms suggest uncomplicated cystitis or recurrent cystitis, the physician may choose to refer the patient to an urgent care facility for a urinalysis and treatment or treat the patient empirically by phone.

"Based on what you've told me you probably have a urinary tract infection. If you wish, I can prescribe an antibiotic over the telephone, or you can be evaluated in an urgent care facility or your primary provider for evaluation. I can also give you something for pain called phenazopyridine, which may turn your urine orange. Drink plenty of fluids, empty your bladder frequently, and follow up with your doctor if your symptoms don't resolve with treatment or if you develop a fever, nausea, vomiting, or abdominal or back pain."

WHAT TO DOCUMENT

Document the patient's symptoms, the presence or absence of any symptoms of more serious infection, presence or absence of vaginal discharge, and risk factors for subacute pyelonephritis or complicated UTI. Document the assessment and whether and when in-person evaluation

WOMAN WHO CALLS TO REPORT DYSURIA

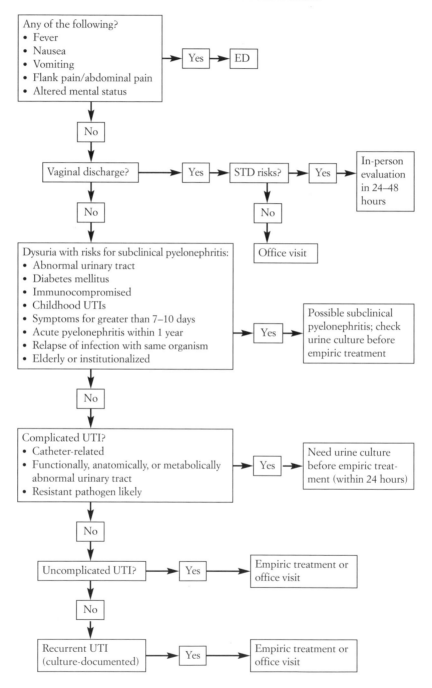

was recommended. If treating empirically, include allergies, other medications, and antibiotic and dose prescribed.

REFERENCES

1. **Komaroff A.** Acute dysuria in adult women. In: Black ER, Bordley DR, Tape TG, Panzer RJ, eds. Diagnostic Strategies for Common Medical Problems. Philadelphia: ACP-ASIM; 1999:243-54.
2. **Stamm WE, Hooton TM.** Management of urinary tract infection in adults. N Engl J Med. 1993;329:1328-34.
3. **Hooton TM, Scholes D, Hughes JD.** A prospective study of risk factors for symptomatic urinary tract infection in young women. N Engl J Med. 1996;335: 468-74.
4. **Stamm WE, Hooton TM.** Urinary tract infections. From pathogenesis to treatment. J Infect Dis. 1989;15:400-6.
5. **Leibovici L, Greenshtain S.** Bacteremia in febrile patients: a clinical model for diagnosis. Arch Int Med. 1991;151:1801-6.
6. **Roberts FJ, Geere IW, Goldman AA.** Three-year study of positive blood cultures. Rev Infect Dis. 1991;13:34-46.
7. **Westrom L.** The effect of acute PID on fertility. Am J Obstet Gyn. 1975;121:707-13.
8. **Westrom L., Joesoef R, Reynolds G.** Pelvic inflammatory disease and infertility. Sex Trans Dis. 1992;19:185-92.
9. **Fairly KF, Bond AG, Brown RB, et al.** Simple test to determine the site of urinary tract infection. Lancet. 1967;2:427.
10. **McCormack W.** Pelvic inflammatory disease. N Engl J Med. 1994;330:115-9.
11. **Leibovici L, Greenshtain S.** Toward empiric management of moderate to severe urinary tract infection. Arch Int Med. 1992;152:2481-6.
12. **Johnson C.** Definitions, classification and clinical presentations of urinary tract infection. Med Clin North Am. 1991;75:241-51.
13. **Komaroff AL, Pass RM, McCue JD.** Management strategies for urinary and vaginal infections. Arch Int Med. 1978;138:1069-73.
14. **Ronald AR, Boutros P, Mourhada H.** Bacteriuria localization and response to single dose treatment in women. JAMA. 1976;235:1854-6.
15. **Fihn SD, Lathan R, Roberts P.** Association between diaphragm use and urinary tract infection. JAMA. 1985;254:240-5.
16. **Strom BL, Collins M, West SL.** Sexual activity, contraceptive use, and other risk factors for symptomatic and asymptomatic bacteriuria. Ann Int Med. 1987; 107:816-23.
17. **Stamm WE, Wagner RF, Amsel R.** Causes of the acute urethral syndrome in women. N Engl J Med. 1980;303:405-9.
18. **Hooten TM, Stamm WE.** Management of acute uncomplicated urinary tract infection. Med Clin North Am. 1991;75:313-25.
19. **Barry HC, Ebell MH.** Evaluation of suspected urinary tract infection in ambulatory women: a cost-utility analysis of office-based strategies. J Fam Pract. 1997;44:49-60.
20. **Saint S, Scholes D, Fihn S.** The effectiveness of a clinical practice guideline for the management of presumed uncomplicated urinary tract infection in women. Am J Med. 1999;106:637-41.

21. **Johnson, JR, Stamm WE.** Diagnosis and treatment of acute urinary tract infections. Infect Dis Clin North Am. 1987;1:773-91.
22. **Hooton TM, Stamm WE.** Overview of acute cystitis, UpToDate, Inc., version 8.1, 1999.
23. **Fowler JE, Pulaski ET.** Excretory urography, cystography and cystoscopy in the evaluation of women with urinary tract infection. N Engl J Med. 1981;304:462-5.
24. **Hooton TM, Stamm WE.** Overview of acute cystitis, UpToDate, Inc., version 7.2, 1999.
25. **Wong ES, McKevitt M.** Management of recurrent urinary tract infections with patient-administered single dose therapy. Ann Int Med. 1985;102:302-7.

9

VAGINAL DISCHARGE

Sally G. Haskell, MD

KEY POINTS

- Women who have dysuria caused by vaginitis usually have vaginal discharge or irritation.
- Women with vaginal discharge who have risk factors for sexually transmitted diseases (STDs) or symptoms suggestive of STDs should be referred for pelvic exam and cultures.
- Women with documented recurrent candidal vaginitis and typical symptoms can be treated empirically.
- Women with less specific typical symptoms of vaginitis at low risk for STDs should be referred to their primary providers on a less urgent basis for pelvic exam and diagnosis.

Vaginal discharge is a common reason for women to seek telephone or in-person advice (1). Telephone evaluation can be crucial in preventing complications, which include complications during pregnancy (caused by trichomoniasis or bacterial vaginosis) as well as pelvic inflammatory disease (PID) and infertility (caused by gonorrhea or chlamydia). The physician should be comfortable with determining the risk for such complications and making appropriate recommendations to the patient over the telephone. He or she should also know which types of vaginitis can be treated empirically.

This chapter will focus on answering the following central questions:

1. Does this patient need urgent or emergency evaluation?
2. If not, when should she be seen in the office?
3. When is empiric treatment indicated?

BACKGROUND

Epidemiology

Vaginal infection or vaginal discharge accounts for 5 to 10 million health care visits per year (2). The most common vaginal infections—bacterial vaginosis, candidiasis, and trichomoniasis—occur at a frequency of 20% to 50%, 20% to 30%, and 5% to 10%, respectively (3). Women complaining of vaginal discharge may also have cervical infection with *Chlamydia trachomatis*, gonorrhea, or herpes simplex.

Utility of Early Diagnosis

The primary purpose of early diagnosis of trichomonas, bacterial vaginosis, or pelvic inflammatory disease is to prevent complications.

Bacterial vaginosis has been associated with an increased risk of preterm birth (4,5), and proper treatment can significantly reduce preterm labor (6). Causal relations have also been established between bacterial vaginosis and pelvic inflammatory disease, plasma cell endometritis, postpartum fever, post-hysterectomy vaginal cuff cellulitis, and post-abortion infection (4,6,7).

Trichomonas infection has been associated with a high prevalence of other STDs and has been found to facilitate transmission of the human immunodeficiency virus (8). If untreated in pregnancy, trichomonas is associated with premature rupture of the membranes, prematurity, and post-hysterectomy cellulitis (9).

Infection with *Neisseria gonorrhoeae* or *Chlamydia trachomatis* can frequently lead to pelvic inflammatory disease. Recent infection with gonorrhea has been found to be associated with adnexal involvement in about half of women in one study (10) and upper genital infection in six of 20 women infected with both gonorrhea and chlamydia in another study (11). Tubal damage and scarring from PID result in important long-term complications such as chronic pelvic pain (18%), ectopic pregnancy (6 times more frequently), and infertility (8% after one episode, 19.5% after two, and 40% after three or more episodes), as well as increased risk for subsequent episodes of PID (12,13). Short-term complications include peritonitis, perihepatitis, sepsis, and death.

The purpose of early telephone diagnosis of less severe causes of vaginitis is to reduce discomfort and other morbidity and, in certain situations, to avoid the need for an office visit.

INITIAL APPROACH TO TELEPHONE DIAGNOSIS

The initial approach to evaluating a complaint of vaginal discharge should involve distinguishing between cystitis and vaginitis or cervicitis because the urgency for "in-person" evaluation and recommendations for empiric treatment will be different.

Patients with vaginitis usually present with symptoms of vaginal discharge or irritation, odor, pruritis, or dyspareunia. When women with vaginitis present with dysuria, they are usually able to sense that it is an external discomfort rather than the internal discomfort located in the urethra and bladder that occurs with urinary tract infections (UTIs) (14-16). Symptoms of vaginitis usually develop weeks to months after exposure in contrast to the more rapid development of dysuria from cystitis or urethritis (17). Chapter 8 discusses the telephone assessment of dysuria in women.

Does the Patient Require an Urgent or Emergency Evaluation?

All pregnant patients with suspected vaginitis or STD should be seen within 24 hours for evaluation and treatment. Although there is no specific data available regarding the potential for increased risk with delay in treatment, it is suggested that these patients be seen within 24 hours because of the possible morbidity associated with untreated infection.

Table 9-1 lists symptoms suggestive of STDs and/or PID. If PID is suspected, the patient, pregnant or not, should also be seen within 24 hours

Table 9-1 Symptoms Suggestive of STDs and/or PID

STD (Not Necessarily PID)	*STD with Suspected PID*
• Mucopurulent vaginal discharge	• Mucopurulent vaginal discharge
• Dysuria	• Dysuria
	• Abdominal pain
	• Fever
	• Nausea, vomiting
	• Dyspareunia

Table 9-2 Risk Factors for Sexually Transmitted Diseases

General Risk Factors
• Young age (highest rates ages 15–30)
• Multiple sexual partners in the past month
• History of sexually transmitted disease
• Sexual contact with an infected partner

Risk Factors for Chlamydia Urethritis
• New sexual partner
• Stuttering onset of symptoms

Risk Factors for Gonoccocal Urethritis
• History of gonorrheal infection
• Indigent inner-city women

Risk Factors for Herpes Simplex Cervicitis
• Most commonly occurs in educated white patients
• Usually occurs within 6 days of contact with an infected partner

because of the significant morbidity and long-term consequences associated with untreated infection. Again, there is no specific data available about the consequences of delayed treatment.

Non-pregnant patients with symptoms of vaginitis should have symptoms and risk factors for PID and STDs assessed (Tables 9-1 and 9-2). Patients with suspected STDs, without symptoms of PID, should be evaluated within 24 to 48 hours (Table 9-3). Patients without risk factors for PID or STDs may be evaluated less urgently or possibly treated empirically.

Trichomoniasis and herpes simplex infection may be difficult to distinguish from other STDs over the phone. Women with trichomonal infection

Table 9-3   Recommended Time Course for Evaluation by Suspected Diagnosis

Suspected Diagnosis	When to Evaluate
PID	Emergently
STD without PID	Urgently (24–48 hours)
Vaginitis	Urgently, semiurgently, or phone sufficient
Recurrent *Candida*	Consider phone evaluation sufficient (empiric)

Table 9-4   Typical Symptoms of Common Vaginal Infections

Infection	Typical Symptoms
Trichomonas	Copious, frothy vaginal discharge; vulvar pruritis, vaginal burning, spotting, dysuria, frequency, urgency
Herpes simplex virus	Painful lesions associated with dysuria, vaginal discharge, tender inguinal lympadenopathy
Candida	Intense vaginal itching or burning; thick whitish curd-like discharge
Bacterial vaginosis	Thin, malodorous vaginal discharge, vulvar itching or burning
Atrophic vaginitis	Vaginal soreness, post-coital burning, dyspareunia, occasional spotting, occasional serosanguinous or watery discharge

generally present with a copious and often frothy vaginal discharge accompanied by vulvar pruritis. Additional symptoms can include vaginal burning, spotting, and symptoms of urethral irritation: dysuria, frequency, and urgency (Table 9-4) (18). The diagnosis is made by saline wet prep of vaginal discharge.

Herpes simplex is seen in 1% to 3% of lower genital tract infections in office practices, with a much higher prevalence in STD clinics. It produces characteristic painful vesicular lesions, which can be associated with dysuria, vaginal discharge, and tender inguinal lymphadenopathy within 6 days of contact with an infected partner (Table 9-4) (19).

Women who call with symptoms of trichomoniasis or herpes simplex should probably be seen within 24 to 48 hours because the physician may be unable, over the phone, to distinguish these symptoms and risk factors from STDs that require more urgent treatment.

If the Patient Does Not Require an Urgent or Emergency Evaluation, When Should He or She Be Seen in the Office?

The patient who calls with symptoms of vaginitis not suspicious for STD or PID should usually be referred for a routine office visit for diagnosis. These symptoms often include a thin, malodorous vaginal discharge with itching or burning (bacterial vaginosis), a curd-like discharge with intense itching or burning (candidiasis), or vaginal dryness or a watery discharge with soreness, post-coital burning, and occasional spotting (atrophic vaginitis) (Table 9-4).

Bacterial vaginosis represents a change in the vaginal ecosystem characterized by a reduction in the prevalence of hydrogen peroxide–producing lactobacilli and an increase in the prevalence of *Gardnerella vaginalis*, *Mobiluncus* species, *Mycoplasma hominis*, anaerobic gram-negative rods and bacteroides, and peptostreptococcus species (20). Patients with this infection usually complain of a thin, malodorous vaginal discharge with accompanying vulvar itching or burning. The vaginal mucosa and vulva may be mildly inflamed, and the odor is usually described as fishy or musty (21).

Candida vaginitis is also very common. One half of women by the age of 25 will have had at least one episode of vulvovaginal candidiasis (22). Sporadic attacks usually occur without a precipitating factor except in women with diabetes mellitis. Risk factors for *Candida* vaginitis include diabetes mellitis, oral contraceptive use, IUD use, recent antibiotic use, and vaginal sponge use (23). Symptoms may include intense vaginal itching or burning associated with a thick curd-like discharge.

Post-menopausal women who are not on estrogen replacement are at risk for atrophic vaginitis. Symptoms of advanced atrophy can include vaginal soreness, post-coital burning, dyspareunia, and occasional spotting. The vaginal mucosa appears thin, with diffuse redness. There may be a serosanguinous or watery discharge. Treatment consists of topical vaginal estrogen.

When Is Empiric Treatment Indicated?

There is some data to support empiric treatment for *Candida* vaginitis in some patients, but further and larger trials are needed. What sounds like typical *Candida* vaginitis over the phone may still be a different type of vaginitis. The commonly taught clinical picture of vaginal or vulvar itching, clumpy white discharge, and lack of odor was found in several studies to be neither sensitive nor specific for *Candida* vaginitis (24,25).

However, one small study identified two clinical criteria that may be useful over the phone: absence of a watery discharge and patient self-diagnosis of "another yeast infection." Each was found to be an independent predictor of a positive culture for *Candida* in women with vaginal discharge.

Empiric antibiotic treatment can be considered in patients with recurrent documented *Candida* vaginitis with similar symptoms. One study found that self-administered empiric treatment of recurrent vulvovaginal candidal infection was preferable and more cost effective than monthly prophylactic treatment (26).

There is no data supporting empiric telephone treatment of patients with vaginal discharge and risk factors for *Candida* vaginitis, such as recent antibiotic use or diabetes mellitus, although such risk factors may help the telephone physician determine the likelihood of infection. There is no data to support empiric telephone treatment of other causes of vaginitis.

In summary, neither the history nor the physical exam has been shown to be reliable in making a diagnosis of *Candida* vaginitis, although data from a small study suggests that patient self-diagnosis of "another yeast infection" and absence of a watery discharge may be helpful. Positive risk factors for candida may help the telephone physician determine the likelihood of infection over the telephone, but there is no data supporting empiric treatment without exam. Patients treated empirically over the telephone should be instructed to call back if they have no response to treatment within 3 days or if any new symptoms develop.

How Does One Advise the Patient About Which Over-the-Counter Antimycotic to Buy?

Topical therapies available over-the-counter for *Candida* vaginitis include nystatin, miconazole, clotrimazole, butaconazole, and ticonazole. Generally, 3-day and 7-day regimens are recommended; 1-day regimens may be less effective (27). A single oral dose of fluconazole, available by prescription, is as effective as 7 days of clotrimazole or miconazole, and may be preferred by some patients, but multiple drug interactions exist, and fluconazole should not be given to any patient with pre-existing liver disease (28). Women with severe inflammation or host factors such as un-controlled diabetes or immunosupression may require 10 to 14 days of topical treatment (28).

WHAT TO TELL THE PATIENT

High Risk: Patient Needs Emergency Evaluation

"You may have a serious infection that could be life-threatening or lead to long-term consequences if untreated. [Depending on the day and time], you should [go to the emergency room/come to the office today/see your gynecologist today/go to an STD clinic today]. You will be examined, have cultures taken, and start antibiotics."

Moderate Risk: Patient May Need Urgent Evaluation (Within 24-48 Hours)

"I am concerned that you may have a serious infection. This should be evaluated in person within the next 2 days. Please call back if you develop fever, abdominal pain, nausea, vomiting, bleeding, or pain with intercourse."

Low Risk: Patient May Schedule a Routine Office Visit (Within 1 Week)

"Based on what you've told me, you probably do not have a serious infec-tion that requires immediate evaluation. You should make an appointment to see your regular doctor this week."

• *For symptoms suggestive of a recurrent yeast infection:* "It sounds very likely that you have a yeast infection. If you are comfortable treating your-self, without being evaluated in person, you can purchase a treatment for yeast infection at your pharmacy such as miconazole, clotrimazole, buta-conazole, or ticonazole, and see your physician soon if your symptoms don't resolve. Without examining you I can't be sure, so if you develop fever, abdominal pain, or other changes in your symptoms, call me back. Call back also if there is no improvement in your symptoms in 3 days."

PATIENT WHO CALLS WITH VAGINAL DISCHARGE

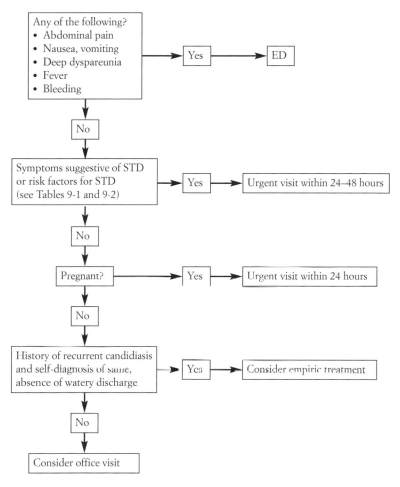

WHAT TO DOCUMENT

Document the presence or absence of symptoms of and risk factors for PID and STDs, the patient's pregnancy status, and her last menstrual period. It is important as well to document your recommendations to the patient (when and where to follow up, when to call back, and what treatments were recommended). Finally, document the patient's understanding of the plan.

REFERENCES

1. **Cullins V, Huggins G.** Non-malignant vulvovaginal disorders. In: Barker LR, Burton JR, Zieve P, eds. Principles of Ambulatory Medicine. Philadelphia: Williams and Wilkins; 1991.
2. **Bickley LS.** Acute vaginitis. In: Black ER, Bordley DR, Tape TG, Panzer RJ, eds. Diagnostic Strategies for Common Medical Problems. Philadelphia: Am Coll Phys; 1999:255-68.
3. **Ferris DG, Hendricks J.** Office laboratory diagnosis of vaginitis. J Fam Pract. 1995;41:575-81.
4. **Oleen-Burkey MA, Hiller SL.** Pregnancy complications associated with bacterial vaginosis. Infect Dis Obstet Gyn. 1995;3:149-57.
5. **Hiller SL, Martius J, Krohn AA.** Case-control study of chorioamniotic infection and histologic chorioamnionitis in prematurity. N Engl J Med. 1988;319:972-8.
6. **McGregor JA, French JI, Parker R, et al.** Prevention of premature birth by screening and treatment of common genital tract infections: results of a prospective controlled evaluation. Am J Obstet Gyn. 1995;173:157-67.
7. **MacDermott R.** Bacterial vaginosis. Br J Obstet Gyn. 1995;102:92-4.
8. **Laga M, Manoka AT, Kivuvu M, et al.** Nonulcerative STDs as risk factors for HIV-1 transmission in women: result of a cohort study. AIDS. 1993;7:95-102.
9. **Soper DE, Bump RC.** Bacterial vaginosis and trichomonas are risk factors for cuff cellulitis after abdominal hysterectomy. Am J Obstet Gyn. 1990;163:1016-21.
10. **Platt R, Rice PA, McCormack WH.** Risk of acquiring gonorrhea and prevalence of abnormal adnexal findings among women recently exposed to gonorrhea. JAMA. 1983;250:3205-9.
11. **Stamm WE, Guinan ME, Johnson C, et al.** Effect of treatment regimens for *Neisseria gonorrhea* on simultaneous infection with chlamydia. N Engl J Med. 1984;310:545-9.
12. **Westrom L.** Effect of acute pelvic inflammatory disease on fertility. Am J Obstet Gyn. 1975;121:707-13.
13. **Westrom L, Joesoef R, Reynolds G.** Pelvic inflammatory disease and infertility. Sex Trans Dis. 1992;19:185-92.
14. **Komaroff AL, Pass TM, McCue JD, et al.** Management strategies for urinary and vaginal infections. Arch Intern Med. 1978;138:1069-73.
15. **Demetriou E, Emans SJ, Masland RP.** Dysuria in adolescent girls: urinary tract infection or vaginitis? Pediatrics. 1982; 70:299-301.
16. **Dans PE, Klaus B.** Dysuria in women. Johns Hopkins Med J. 1976;138:13-8.
17. **Komaroff A.** Acute dysuria in adult women. In: Black ER, Bordley DR, Tape TG, Panzer RJ, eds. Diagnostic Strategies for Common Medical Problems. Philadelphia: Am Coll Phys. 1999:243-54.
18. **Anderson JR.** Genital tract infection in women. Med Clin North Am. 1995; 79;6:1271-98.
19. **Corey L.** Genital herpes. In: Holmes K, Mardh P, eds. Sexually Transmitted Diseases. New York; McGraw-Hill: 1984;449-74.
20. **Sobel JD.** Vaginitis. N Engl J Med. 1997;26:896-903.
21. **Amsel R, Toffen PA.** Nonspecific vaginitis diagnostic criteria. Am J Med. 1983;74:14-22.

22. **Geiger AM.** The epidemiology of vulvovaginal candidiasis among university students. Am J Pub Health. 1995;85:1146-8.

23. **Foxman B.** The epidemiology of vulvovaginal candidiasis. Am J Pub Health. 1990;80:329-31.

24. **Abbot J.** Clinical and microscopic diagnosis of vaginal yeast infections. Ann Emerg Med. 1995;25:587-91.

25. **Sweet RI.** Importance of differential diagnosis in acute vaginitis. Am J Obstet Gyn. 1985;152:921-3.

26. **Fong IW.** The value of prophylactic (monthly) clotrimazole versus empiric self-treatment in recurrent vaginal candidiasis. Genitourinary Med. 1994;70:124-126.

27. Medical Letter. Vol. 41 (issue 1062); September 24, 1999.

28. **Sobel JD, Brooker D, Stein GE.** Single dose oral fluconazole compared with conventional clotrimazole topical treatment for *Candida* vaginitis. Am J Obstet Gynecol. 1995;172:1263.

10

SORE THROAT

Timothy S. Loo, MD

KEY POINTS

- There is no consensus on an optimal strategy to evaluate patients with sore throat either in an ambulatory setting or over the telephone
- Most sore throats are caused by viral agents, are self-limited, and do not benefit from antibiotic therapy. However, pharyngitis caused by group A streptococcus can have serious suppurative and nonsuppurative sequelae that can be prevented with antibiotic therapy.
- Streptococcal pharyngitis classically presents with an acute onset of sore throat, fever, tonsillar exudate, and tender cervical adenopathy. However, these clinical symptoms and signs are not specific enough to reliably distinguish streptococcal from non-streptococcal pharyngitis
- Viral pharyngitis can sometimes be distinguished by the prominence of rhinorrhea, cough, hoarseness, conjunctivitis, or diarrhea, which are not typical of streptococcal pharyngitis.
- Because clinical diagnosis is unreliable, diagnosis depends on laboratory testing. Consequently, a patient presenting with symptoms suspicious for streptococcal pharyngitis should present to a medical facility for a rapid streptococcal antigen test or streptococcal culture, and the decision to treat should be based on a positive test result.
- Any patient whose sore throat is complicated by serious respiratory compromise manifested by stridor, dyspnea, or tachypnea should be directed immediately to the nearest emergency department.

Sore throat is one of the most common reasons patients seek medical attention. It is most often benign with a self-limited course, but sore throat can also be the presenting symptom of a more serious infection that could result in local suppurative complications, systemic complications (rheumatic fever, glomerulonephritis, septic arthritis), and severe respiratory compromise. Telephone evaluation should distinguish those patients with benign self-limited sore throats from those who require further evaluation and possibly urgent treatment.

This chapter will focus on the following questions:

1. Which patients require emergency evaluation?
2. How do the different causes of pharyngitis present? Can streptococcal and viral pharyngitis be reliably diagnosed over the phone?
3. How does one manage suspected or confirmed streptococcal pharyngitis over the phone?
4. How does one treat viral pharyngitis symptomatically?

BACKGROUND

Epidemiology

Most sore throats are caused by viral agents. It is estimated that only about 15% of sore throats are caused by group A streptococcus (1). Common pathogens causing pharyngitis include group A and G streptococci, *Chlamydia pneumoniae*, *Mycoplasma pneumoniae*, adenovirus, influenza A and B, and parainfluenza viruses. Less common pathogens include *Corynebacterium diphtheriae, Neisseria gonorrhoeae, Arcanobacterium haemolyticus,* Epstein-Barr virus, herpes simplex virus, coxsackievirus A, and human immunodeficiency virus (Table 10-1). No etiologic agent is detected in about one-half of cases (2). Post-nasal drip, irritant exposure, and lack of humidity might account for some of these cases.

Streptococcal pharyngitis can occur at any age but is most common among school-age children. The majority of cases occur in the fall, winter, and early spring. Transmission follows contact with respiratory secretions.

Table 10-1   Common and Less Common Pathogens Causing Sore Throat

Common Pathogens		Less Common Pathogens	
Name	*Percent of Cases*	*Name*	*Percent of Cases*
Viral	30–60% (6)	*Corynebacterium diptheriae*	…
Group A streptococcus	15% (1)		
Non-group A streptococcus	10–15% (3)	*Neisseria gonorrhoeae*	1% (2a)
Chlamydia pneumoniae	9% (2a)	*Arcanobacterium haemolyticus*	…
Mycoplasma pneumoniae	10–15% (3)	Epstein-Barr virus	1–6% (2a)
Adenovirus	…	Herpes simplex virus	…
Influenza	…	Coxsackievirus A	…
Parainfluenza	…	Human immuno-deficiency virus	…

Close person-to-person contact in schools, child care settings, and military installations facilitate transmission and cause clustering of cases.

Utility of Early Diagnosis and Treatment

Early identification and treatment of the infections causing sore throat can prevent significant morbidity and mortality. Although uncommon, epiglottitis and pharyngitis caused by *Corynebacterium diptheriae* can cause severe respiratory symptoms and death if not quickly recognized and treated. *Neisseria gonorrhoeae,* another uncommon cause of pharyngitis, may result in septic arthritis if not treated in a timely manner. Group A streptococcal pharyngitis can cause serious suppurative (such as peritonsillar cellulitis, peritonsillar abscess, retropharyngeal abscess, otitis media, sinusitis, and suppurative cervical adenitis) and nonsuppurative complications (acute rheumatic fever and acute glomerulonephritis). Although timely antibiotic therapy can prevent acute rheumatic fever and suppurative complications, antibiotic therapy is not thought to prevent acute glomerulonephritis.

A subset of patients with viral pharyngitis can sometimes be distinguished. Because these cases are usually self-limited and do not benefit from antibiotic therapy, their identification may obviate the need for urgent clinical evaluation and prevent the inappropriate use of antibiotics.

DOES THE PATIENT REQUIRE EMERGENCY EVALUATION?

Generally, the only patients with sore throat that require immediate evaluation are those at risk for airway obstruction. Although these patients make up a small minority of cases, it is critical that they not be missed. The agents responsible for these cases include *Corynebacterium diptheriae, Haemophilus influenzae, Streptococcus pneumoniae* and *pyogenes,* and *Staphylococcus aureus.* These patients might complain of odynophagia, hoarseness, or dyspnea. Stridor or tachypnea may be evident over the telephone or be described by the companion calling. Given the life-threatening consequences if left untreated, patients with these symptoms or signs should proceed immediately to the nearest emergency room for more detailed evaluation.

WHICH PATIENTS ARE AT RISK FOR COMPLICATIONS?

Among patients who do not need immmediate evaluation, the only common cause of pharyngitis causing serious complications is group A streptococcus. Complications of group A streptococcal pharyngitis include local suppurative complications and systemic complications (e.g., rheumatic fever and glomerulonephritis). These complications, with the exception of glomerulonephritis, are preventable with antibiotic therapy, making early diagnosis and treatment imperative. In contrast, non-streptococcal pharyngitis is self-limited, does not require further evaluation, and has not been shown to benefit from antimicrobial therapy. Consequently, evaluation focuses on distinguishing streptococcal from non-streptococcal pharyngitis.

In addition, *Neisseria gonorrhoeae,* a rare cause of pharyngitis (occurring in approximately 1% of cases), can cause septic arthritis if disseminated. As already mentioned, severe respiratory compromise is another rare complication of infections associated with pharyngitis.

HOW DO THE DIFFERENT CAUSES OF PHARYNGITIS PRESENT?

The various etiologic agents causing pharyngitis present with many of the same symptoms and signs, making clinical differentiation difficult. There are certain characteristics or typical presentations that can suggest a particular etiologic agent. Although these features can be helpful, it is important to recognize that classic or typical presentations represent the minority of cases.

Viral Pharyngitis

Viral pharyngitis can sometimes be distinguished by the prominence of various symptoms and signs that are simply not typical of streptococcus pharyngitis. These symptoms include coryza, hoarseness, cough, and diarrhea. Signs that a patient on the phone may be able to describe include conjunctivitis, anterior stomatitis, and discrete ulcerative lesions in the oropharynx.

Group A Streptococcal Pharyngitis

Group A streptococcal pharyngitis typically presents with the acute onset of sore throat and fever with a temperature greater than 100°F (37.8°C). Patients will often complain of pain on swallowing and a sensation of swollen glands. Associated systemic symptoms include headache, malaise, nausea, and anorexia. The symptoms should have a duration of less than 1 week. Associated signs that some patients may be able to describe over the telephone include tonsillopharyngeal erythema and yellow-gray exudate, soft palate petechiae, beefy red swollen uvula, tender anterior cervical adenopathy, and scarlatiniform rash.

Other Infectious Agents

Pharyngitis caused by *Mycoplasma pneumoniae* and *Chlamydia pneumoniae* are more likely to have cough and other symptoms of a lower respiratory infection. Gonococcal pharyngitis typically presents (>95% cases) with concurrent symptoms of urethritis or vaginitis (3). It should be considered in those practicing orogenital sex and those with persistent sore throat despite treatment with penicillin. Infectious mononucleosis typically occurs in adolescents or young adults and usually presents with malaise, an exudative pharyngitis, marked lymphadenopathy, and splenomegaly. *Corynebacterium diptheriae* presents with a grayish membrane involving the anterior nares, tonsils, uvula, or pharynx. Pathogens causing epiglottitis can produce an enlarged inflamed epiglottis that protrudes into the hypopharynx. Both *Corynebacterium diptheriae* and pathogens causing epiglottitis can cause partial or complete airway obstruction that may manifest as stridor or tachypnea.

CAN STREPTOCOCCAL PHARYNGITIS BE RELIABLY DIAGNOSED OVER THE TELEPHONE?

Unfortunately, clinical findings alone do not adequately distinguish streptococcal from non-streptococcal pharyngitis. Although typical features of streptococcal pharyngitis such as acute onset of fever, tonsillar exudate, and tender adenopathy are supportive of a diagnosis of streptococcus pharyngitis, they are not pathognomonic and are estimated to be present in only about 15% of patients with streptococcal pharyngitis (4).

Numerous studies have demonstrated that clinical diagnosis of streptococcal pharyngitis is unreliable. A recent review of studies comparing clinical diagnosis with throat culture in primary care practices found that clinicians accurately predicted positive throat cultures in only 45% to 72% of patients and negative throat cultures in 57% to 80% of patients (5). In other words, clinicians relying on clinical impression alone would fail to treat one quarter to one half of patients with positive cultures and would needlessly treat one fifth to one half of patients with negative cultures. Because the majority of cases are not caused by group A streptococcus, the net result would be a significant overuse of antibiotics.

Scoring systems (6-8) based on combinations of findings have been devised in an attempt to improve diagnostic accuracy. These scoring systems have not found wide acceptance because of their complexity, lack of validation in different populations, and an inferior ability to discriminate streptococcal compared to throat culture. The degree to which sensitivity is sacrificed to improve specificity or to which specificity is sacrificed to improve sensitivity limits their practical utility. In addition, the scoring systems studied are not directly applicable to telephone evaluation because they are dependent on physical findings often in combination with laboratory results, data not available to the clinician evaluating a patient over the telephone.

Can Viral Pharyngitis Be Reliably Diagnosed Over the Telephone?

Although a number of the symptoms and signs of viral pharyngitis are distinct from those of streptococcal infection, there is little published evidence that they can be used to reliably distinguish between the two. Despite this, the American Academy of Pediatrics suggests that those presenting with these particular symptoms or signs are at very low risk of having streptococcal pharyngitis and should not undergo further testing (9). Because untreated streptococcal pharyngitis has the potential for serious complications, any patients with equivocal presentations should undergo diagnostic testing to rule it out (see next section).

Should Antibiotic Treatment Be Prescribed Over the Telephone if Streptococcal Pharyngitis Is Suspected?

Because clinical identification of streptococcal pharyngitis is unreliable, antibiotics should generally not be prescribed over the phone for suspected streptococcal pharyngitis. Consequently, most expert groups, including the CDC and the American Academy of Pediatrics, recommend using throat culture as a basis for management (1,9). Clinicians are advised to wait for a positive rapid streptococcal antigen test or culture before instituting antibiotic therapy. This approach minimizes inappropriate antibiotic administration, antibiotic-related adverse effects, and the development of antibiotic resistance.

Exceptions to this approach include patients with a history of rheumatic fever not currently on antibiotic prophylaxis and new cases of pharyngitis in the setting of an explosive streptococcal epidemic in a semiclosed population. In these cases, antibiotic therapy should be started empirically before culture results are known and then discontinued if culture results are negative (discussed below).

How Good Are Laboratory Tests for Diagnosing Streptococcal Pharyngitis?

Throat Culture

Currently, throat culture is the gold standard for diagnosing streptococcal pharyngitis. False negative cultures occur in less than 10% of symptomatic patients if properly obtained, cultured, and interpreted (9). Although a positive culture does not distinguish patients with acute streptococcal pharyngitis from those who are streptococcus carriers, most experts recommend assuming that a positive culture result is significant and to treat accordingly. This approach results in some degree of overtreatment, but far less than if clinicians were to treat on clinical impression alone. Culture results are usually available within 24 hours but held for an additional 24 hours if negative to maximize yield.

Rapid Streptococcal Antigen Testing

Rapid streptococcal antigen testing detects a carbohydrate antigen by either agglutination or enzyme immunoassay. Antigen testing offers two distinct advantages over throat culture: rapid results (within an hour) and the ability to perform the test in the office. Rapid antigen tests have a reported sensitivity between 76% and 87% and a specificity between 90% and 96% (6). Because of the high specificity of antigen testing, a positive result does not require culture confirmation and is an adequate basis to initiate treatment. When the result is negative, however, a confirmatory culture should be performed because of the limited sensitivity of the test.

Does the Risk for Rheumatic Fever Increase by Delaying Treatment?

Delaying treatment for a few days while awaiting culture results does not increase the risk for rheumatic fever. Antibiotic therapy as long as 9 days after the onset of acute illness is still effective in preventing rheumatic fever (9). There is even some suggestion that delaying treatment for 48 hours can decrease the risk of a recurrence, possibly because early treatment inhibits the development of specific antibody (4).

Recommendations differ, however, for those with a previous history of rheumatic fever. For patients with a history of rheumatic fever who are no longer on antibiotic prophylaxis, empiric antibiotic therapy should be started at presentation and discontinued only if tests return negative. For the same reason, it is recommended that the threshold to culture be lower in these patients.

What Treatment Should Be Recommended for Patients with Confirmed Streptococcal Pharyngitis?

If called by the laboratory with either a positive streptococcal antigen or culture, initiate antimicrobial treatment with:

- Penicillin V 250 mg orally three times a day for 10 days, *or*
- Benzathine penicillin G 1.2 million U IM once if compliance is questionable, *or*
- Erythromycin orally for 10 days if penicillin-allergic (estolate 20 to 40 mg/kg per day in two to four divided doses or succinate 40 mg/kg per day in two to four divided doses)

WHAT TO TELL THE PATIENT

Possible Viral Pharyngitis

"Your sore throat is most likely caused by a virus. Although it can make you uncomfortable, no antibiotic has been shown to be helpful for this kind of infection. Use over-the-counter medications to treat the symptoms. Use decongestants for nasal stuffiness, throat lozenges and cough suppressants for persistent cough, anti-pyretics such as acetaminophen for fever, analgesics for body aches, and anti-diarrheals for loose stools. If symptoms change dramatically or are not resolving, call your doctor."

Possible Streptococcal Pharyngitis

"Your sore throat could be caused by an infection that could have serious consequences if not treated. Without treatment, this infection could

PATIENT WHO CALLS WITH SORE THROAT

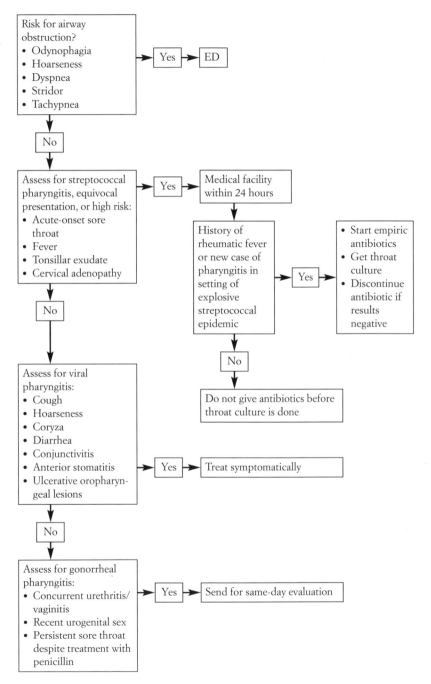

spread to other areas of the throat or could cause a serious condition called rheumatic fever. The only way to know for sure is to have a test done either in a medical clinic or in a nearby urgent care center or emergency department. If the tests are positive, you will be prescribed an antibiotic that will need to be taken for 10 days unless you are given a single intramuscular injection of antibiotic. Serious consequences could still occur if you do not complete the full 10 days of antibiotics. If you are sent home after a culture is taken and no antibiotic is prescribed, make sure to call your regular doctor the next day to find out the result of the test."

WHAT TO DOCUMENT

Document the patient's symptoms and the presence or absence of risks for airway obstruction or gonorrheal pharyngitis, history of rheumatic fever, and level of suspicion for streptococcal pharyngitis. Include your recommendations to the patient in terms of disposition and symptomatic medications.

REFERENCES

1. Pharyngitis. CDC Web Site: www.cdc.gov.
2. **Bisno A.** Streptococcus Pyogenes. In: Mandell G, Bennett J, Dolin R, Eds. Principles and Practice of Infectious Diseases, 4th ed. New York: Churchill Livingstone; 1995:1786-99.
2a. **Komaroff A.** Sore throat and acute infectious mononucleosis in adult patients. In: Strategies for Common Medical Problems, 2nd ed. Philadelphia: American College of Physicians; 199, 29-242.
3. **Koster F.** Respiratory tract infections: Pharyngitis. In: Principles of Ambulatory Medicine, 5th ed. Baltimore: Williams & Wilkins; 1999:346-9.
4. Diagnosis and treatment of streptococcal sore throat. Drug Ther Bull. 1995;33:9-12.
5. **McIsaac WJ, Goel V, Slaughter PM, et al.** Reconsidering sore throats. Part 1. Can Fam Physician. 1997;43:485-93.
6. **Pichichero ME.** Group A streptococcal tonsillopharyngitis: cost effective diagnosis and treatment. Ann Emerg Med. 1995;25:390-402.
7. **McIsaac WJ, Goel V, Slaughter PM, et al.** Reconsidering sore throats. Part 2. Can Fam Physician. 1997;43:495-500.
8. **Pichichero ME, Disney FA, Green JL, et al.** Comparative reliability of clinical, culture, and antigen detection methods for the diagnosis of group A beta-hemolytic streptococcal tonsillopharyngitis. Pediatr Ann. 1992;21:798-805.
9. **American Academy of Pediatrics.** Group A streptococcal infections. In: 1997 Red Book: Report of the Committee on Infectious Diseases, 24th ed. Elk Grove, IL: American Academy of Pediatrics. 1997:483-94.

11

....

HEADACHE

....

Kei Mukohara, MD • David L. Stevens, MD
Mark Schwartz, MD

KEY POINTS

- Evaluation should focus on determining whether the patient's problem is an emergency (requires immediate emergency department evaluation), is urgent (requires in-person office evaluation within 24 hours or a few days), or is nonurgent (requiring reassurance, analgesia, and possibly follow-up to ensure improvement of symptoms).
- Emergency evaluation should be considered when there is suspicion of altered mental status or neurologic deficits, head trauma, headache described as "worst headache ever," and very severe/intolerable symptoms.
- Urgent in-person evaluation should be strongly considered for headache with increasing frequency or severity, new-onset headache in a patient who is 30 years old or older, headache in a patient who has cancer or is HIV-infected, concomitant infection, headache with vigorous exercise, or change in character of usual headache.
- Physicians should avoid making specific diagnoses such as tension headache or migraine over the telephone because these diagnoses require establishment of a normal neurologic exam.
- Patients judged to be low-risk for dangerous pathology can be managed with analgesics and should be counseled about calling back for warning signs of emergency or urgent conditions.

This chapter will discuss when in-person evaluation of a patient with headache is indicated and what treatment strategies are likely to be effective when in-person evaluation is not indicated. It will also address the challenge of assessing risk without a neurologic exam. The determination of a specific diagnosis is not a primary focus of this chapter because diagnosis without physical exam and imaging is often impossible. Moreover, on the telephone, deciding on a specific diagnosis is not essential; diagnosing intracranial hemorrhage versus ischemic stroke or glioblastoma versus toxoplasmosis is not as important as ensuring that the patient who requires further evaluation is seen and imaged within an appropriate time period.

A sensible approach for telephone management of headache is to categorize the headache into one of three groups based on how soon the patient needs to be treated to avoid death or permanent disability. The three levels are: emergency, urgent, and not requiring in-person evaluation.

Table 11-1 lists some of the specific diagnoses included in each of these groups. As stated above, the physician on the telephone does not need to decide between diagnoses within a group, only into which group the patient fits. He or she needs to determine whether the patient requires emergency evaluation or urgent evaluation and how headaches

Table 11-1 Headache Etiologies and the Need for In-Person Evaluation

Emergency Evaluation	*Urgent Evaluation (Within 24 Hours If Possible, At Least Within 2–3 Days)*	*In-person Evaluation (Only If Telephone Management Ineffective or Symptoms Change)*
• Subarachnoid hemorrhage	• Brain tumor without obvious neurologic impairment	• Migraine
• Intracranial hemorrhage		• Tension-type headache
• Stroke	• Chronic brain infection, such as cryptococcal meningitis in HIV-infected persons	
• Acute meningitis		
• Traumatic head injury		
• Unbearable pain regardless of cause		

with a low likelihood of dangerous pathology should be managed over the telephone.

BACKGROUND

Epidemiology

Non-Traumatic Headaches: Prevalence of Intracranial Pathology

Studies from emergency department and office settings show that the rate of intracranial pathology in patients with headache without trauma is less than 5% (Table 11-2). Although there are no studies that directly address the prevalence of serious pathology among patients calling a physician for headache, the studies in Table 11-2 may nevertheless be useful in generating a conservative estimate of the probability of serious disease in a patient calling with headache. The most common diseases requiring emergency attention were subarachnoid hemorrhage (prevalence 1.3% to 1.7%) (1,3) and subdural hematoma (prevalence 1.0%) (1).

Traumatic Headaches: Prevalence of Intracranial Pathology

In contrast to the low risk associated with non-traumatic headache, the risk of serious disease in patients with headache resulting from minor head trauma is quite a bit higher. An emergency department study showed a risk of serious disease to be 23%, with 9% requiring surgery (5). Clearly these patients require a much greater degree of caution.

Table 11-2   Prevalence of Intracranial Pathology in Patients Presenting with Headache with No History of Trauma

Reference	Setting	Number of Patients	Rate of Intracranial Pathology (% [95% Confidence Interval])
1	Primary Care Office	293	2.7 [0.84–3.6]
2	ED	291	4.1 [1.8–6.4]
3	ED	468	3.8 [2.1–5.5]

Utility of Early Diagnosis

Although the probability of intracranial pathology may be low (at least in non-traumatic headaches), the stakes involved remain high because of the gravity of the possible diagnoses. Many of these conditions can be deadly if not treated promptly.

For example, 10% of patients with subarachnoid hemorrhage die before hospitalization (though only 1/3 respond well to treatment), and 50% to 65% of patients with undiagnosed subdural hematoma will die, the number being even higher in the elderly (7). Without the advice of a physician on the telephone, a patient may wait too long before seeking medical help. On the other hand, patients with symptoms consistent with a low risk headache can be reassured and advised about treatment over the telephone, saving a costly trip to the emergency department—an experience that might make the headache worse!

TELEPHONE EVALUATION

Studies performed using in-person evaluations suggest that specific elements of the patient's history can be useful in assessing risk of serious pathology. However, these studies generally assess history and physical exam together.

Physicians may sometimes perform a "physical exam by proxy," asking the patient or companion about motor or other significant neurologic deficits, such as arm or leg weakness or facial droop. Although no evidence exists to support this practice, a patient's report of facial droop or other neurologic deficit clearly raises the probability of an intracranial process. Consequently, normal gait, speech, and arm strength, as observed by the patient or companion, may lower the probability of acute intracranial process, but the patient or family member might not notice more subtle changes.

There are no data, however, to suggest that a physician is on solid ground trusting a layperson's physical exam. It would be prudent to allow a family member's observation of a neurologic deficit to persuade an undecided physician *to recommend* immediate evaluation, but it is inadvisable to rely on a layperson's exam *to rule out* the need for immediate evaluation.

Clearly, the evaluation of headache without a physical exam is difficult. The following discussion shows what characteristics of the patient's history are helpful to determining the necessity of in-person evaluation.

Does the Patient Require Emergency Evaluation?

Studies of patients with headache presenting to an emergency department found that a number of individual historical factors predicted intracranial pathology requiring emergency evaluation *independent* of the neurologic exam (Table 11-3): acute or sudden onset, occipital location, presence of multiple associated symptoms, age equal to or greater than 55 years, altered mental status, and vomiting (2,3). Although none of these alone is powerful enough to be diagnostic, they all raise the probability of intracranial

Table 11-3   Items in Clinical Examination That Are Useful in Diagnosis of Emergency Intracranial Pathology in Patients with Headache

Findings	Positive Likelihood Ratio	Negative Likelihood Ratio
Headaches of acute onset (3)	2.24	0.33
Occipital location (3)	4.74	0.26
Presence of multiple associated symptoms (3)	2.29	0.53
Age 55 or older (3)	2.72	0.5
Abnormal findings on neurologic examination (3)	16.2	0.62
Presence of *any* of the following (2): • Age 60 or older • Focal neurological deficit • Headache with vomiting • Altered mental status	1.5	0.0
"Worst headache of my life" (no studies)	N/A	N/A

* N/A = not applicable.

pathology requiring emergency evaluation and are more highly predictive when present or absent in combination. One author argues for imaging all patients with any of the "high risk" indicators (3). It is conventional wisdom that when a patient experiences "the worst headache of my life," subarachnoid hemorrhage needs to be ruled out. Until data are available to suggest otherwise, it is prudent to refer such patients for emergency evaluation.

Can One Rule Out Meningitis over the Phone?

There is evidence from emergency department studies that meningitis can be ruled out in the absence of all three of the following (sensitivity 99% to 100%): fever, neck stiffness, and altered mental status (4). However, ascertaining information about these symptoms over the phone has significant limitations. First, it is difficult to assess mental status. Asking a family member to confirm mental status may be helpful, but an anxious family member might not perceive subtle mental alterations. Additionally, in this study, temperature and neck stiffness were determined in person by health professionals. Therefore, the apparent absence of fever, neck stiffness, and altered mental status in a patient evaluated over the phone is not as reassuring as it would be after an in-person evaluation. Given the gravity of meningitis and efficacy of treatment, it is prudent to err on the side of in-person evaluation if the history is suggestive of meningitis.

What Is the Risk of Intracranial Injury in Patients Who Report Headache After Minor Head Trauma?

Intracranial injury appears to be much more common in patients with headache resulting from minor head trauma. An emergency department study found that patients with headache after minor head injury had a 23% rate of intracranial injury and a 9% rate of needing surgical intervention (5). The study also showed that the absence of loss of consciousness did not rule out intracranial pathology. Although it is difficult to know to what extent these high incidences apply to patients calling on the phone, it seems prudent to advise all patients calling with headache after head trauma to be seen immediately for further evaluation.

Table 11-4 Predictors of Intracranial Disease Requiring Urgent Evaluation (Within 24 to 48 Hours)

- Subacute headaches with increasing frequency or severity
- New-onset headache after age 30
- New-onset headache in patients who have cancer or HIV
- Change in character of usual headache
- Concomitant infection
- Onset of headache with vigorous exercise

Does the Patient Require Urgent Evaluation (Within 24 to 48 Hours) for Potentially Serious Underlying Disease?

Diagnoses such as tumors and chronic infections may eventually be life threatening but do not require immediate treatment. Patients with these diagnoses can usually wait 24 to 48 hours without increasing their risk of long-term complications.

There are no studies that specifically assessed callers with headaches who have urgent rather than emergency conditions. Expert opinion and the usual standard of care suggest that the criteria listed in Table 11-4 predict increased risk of intracranial pathologies (6,7). It is advisable for callers with headache who have any of the features listed in Table 11-4 to be seen by primary care physicians or other available providers within 24-48 hours.

How Should Patients Not Needing Emergency or Urgent Evaluation Be Managed?

Two common, less serious types of headache are migraine and tension-type headache. Sinusitis, another cause of headache, is addressed specifically in Chapter 7. Patients with a history of migraine or tension-type headaches may call for an exacerbation or medication refill. In this case, physicians should review the patient's symptoms and compare them to the International Headache Society criteria for diagnosis of migraine and tension headaches (Tables 11-5 and 11-6). If the symptoms are consistent with International Headache Society criteria (see Tables 11-5 and 11-6) (8,9), the physician should consider recommending and/or

---

Table 11-5   International Headache Society Diagnostic Criteria for Migraine

- Headache attacks last 4-72 hours

- Headache has at least two of the following characteristics:
  —Unilateral location
  —Pulsating quality
  —Moderate or severe intensity
  —Aggravation by routine physical activity

- During headache, at least one of the following occurs:
  —Nausea and/or vomiting
  —Photophobia and phonophobia

- At least five attacks fulfilling above criteria

- History, physical examination, and neurologic examination do not suggest any underlying organic disease

- Patients may or may not present with prodromal aura

Modified from Reference 9.

Table 11-6   International Headache Society Diagnostic Criteria for Tension-Type Headache (Headache Lasting 30 Minutes to 7 Days)

- Headache has at least two of the following characteristics:
  —Pressing/tightening quality
  —Mild or moderate intensity
  —Bilateral location
  —No aggravation by routine physical activity

- No nausea or vomiting with headache

- Photophobia and phonophobia are absent, or one but not the other is present

- History, physical examination, and neurologic examination do not suggest any underlying organic disease

---

Modified from Reference 9.

---

prescribing whatever medications have been effective for the patient in the past.

On the other hand, physicians should refrain from making a specific diagnosis in patients with no history of migraine or tension-type headaches. The physician should first assess risk for dangerous pathology as outlined above. If the patient has a low risk for dangerous pathology, the headache should be managed with NSAIDs such as ibuprofen or naproxen. The physician should refrain from making a specific diagnosis for the following reasons:

1. International Headache Society criteria for diagnosis of migraine and tension-type headaches require physical and neurologic examination.
2. NSAIDs, and probably acetaminophen to a lesser degree, are effective in alleviating both tension-type headaches (11) and migraine attacks of moderate intensity (8).

MANAGEMENT OF NON-DANGEROUS HEADACHES

In managing non-dangerous headaches over the phone, it is advisable that physicians recommend non-specific analgesics, such as NSAIDs, for the reasons mentioned above. A possible exception is refilling migraine-abortive medication (such as sumatriptan) for a recurrence of the patient's typical migraine symptoms.

Nonsteroidal Anti-Inflammatory Drugs

NSAIDs have been shown in randomized trials to be superior to placebo in relieving symptoms in non-dangerous headaches. For example, naproxen (750 mg initially, followed by 250-500 mg as needed up to 1250 mg per 24 hours) and ibuprofen (initial dose 1200 mg) have been shown to reduce migraine duration and severity and nausea (7). Indomethacin is also a useful abortive treatment for migraine, especially in suppository form, for patients unable to take oral medications during an attack (7).

In tension headaches, a randomized trial showed ibuprofen (400 mg or 800 mg) to be more effective than aspirin 650 mg and placebo (11), whereas naproxen sodium (275 mg to 550 mg) has been shown to be superior to acetaminophen and placebo in relieving pain (12,13).

Acetaminophen (With or Without Caffeine)

Acetaminophen, despite its better side effect profile, is often less effective for non-dangerous headaches. The study mentioned above showed its inferiority to the NSAID naproxen in tension headaches. An exception to this is acetaminophen in combination with aspirin and caffeine (Excedrin, 2 extra-strength tablets) for migraines. This combination was shown to be effective in relieving migraine compared with placebo in a randomized controlled trial; 59% of patients treated with Excedrin had headache intensity of mild to none 2 hours after administration compared with 33% of those treated with placebo (10).

Patient Self-Monitoring

Self-monitoring is an essential component of management of what seems to be a non-dangerous headache. What initially appears to be a benign headache may progress to a dangerous one. Patients should be advised to call back if they develop new symptoms. They should see their primary care physician within a week if the headache does not improve with analgesics (see What to Tell the Patient).

WHAT TO TELL THE PATIENT

Patient Needs Emergency Evaluation

"You may be having a stroke [or whatever tentative diagnosis] which can be life-threatening. You should call 911 now and tell them you are having a very severe headache; they will take you to an emergency department. In the emergency department, they will examine you carefully and may perform a CT scan or other tests to determine if this is something serious and to see what treatment you need." The physician should place the call for an ambulance if the patient is alone, has a change in mental status or has rapidly progressing symptoms.

Patient Needs Urgent (Within 24-48 Hours) Evaluation

"Based on what you've told me, you don't need to go to the emergency department right now. However, I do think we should evaluate this headache further because I am concerned that there's a chance this headache may be caused by something serious. You should see me or another available

PATIENT WHO CALLS WITH HEADACHE

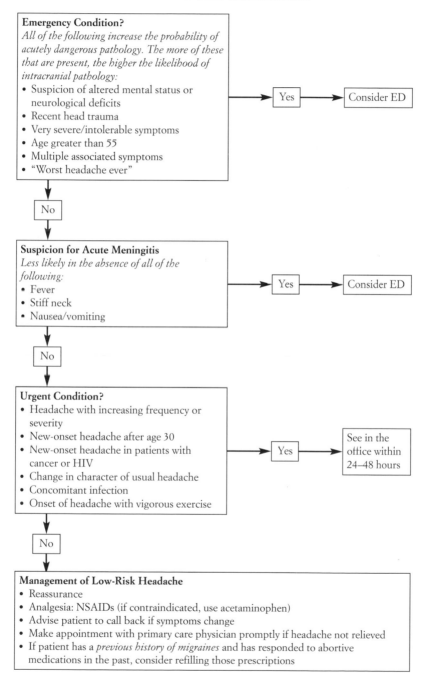

Emergency Condition?
All of the following increase the probability of acutely dangerous pathology. The more of these that are present, the higher the likelihood of intracranial pathology:
- Suspicion of altered mental status or neurological deficits
- Recent head trauma
- Very severe/intolerable symptoms
- Age greater than 55
- Multiple associated symptoms
- "Worst headache ever"

Yes → Consider ED

No

Suspicion for Acute Meningitis
Less likely in the absence of all of the following:
- Fever
- Stiff neck
- Nausea/vomiting

Yes → Consider ED

No

Urgent Condition?
- Headache with increasing frequency or severity
- New-onset headache after age 30
- New-onset headache in patients with cancer or HIV
- Change in character of usual headache
- Concomitant infection
- Onset of headache with vigorous exercise

Yes → See in the office within 24–48 hours

No

Management of Low-Risk Headache
- Reassurance
- Analgesia: NSAIDs (if contraindicated, use acetaminophen)
- Advise patient to call back if symptoms change
- Make appointment with primary care physician promptly if headache not relieved
- If patient has a *previous history of migraines* and has responded to abortive medications in the past, consider refilling those prescriptions

health care provider in the office tomorrow or within the next few days. In the meantime, take [NSAIDs, acetaminophen]. If the headache changes at all, or if you develop new symptoms such as fever, stiff neck, vomiting, or weakness in any part of your body or anything else that worries you, be sure to call me back right away, or call 911."

Patient Does Not Need In-Person Evaluation

"Based on what you've told me, your headache is not likely to be caused by something dangerous. The best thing to do now is to try some pain relievers such as [ibuprofen, etc.]. If the headache changes at all, or if you develop new symptoms such as fever, stiff neck, vomiting or weakness in any part of your body or anything else that worries you, be sure to call me back right away. If the headache doesn't seem to be getting better, call back or see your primary care physician within a week."

WHAT TO DOCUMENT

- Specific symptoms described, including the severity of the pain, any associated symptoms, whether the headache is of new onset, the presence or absence of recent head trauma, fever, stiff neck, nausea, or vomiting.
- Relevant past medical history such as cancer or HIV.
- Patient's mental status.
- Patient's age.
- For patients with a history of headache, whether the current headache is similar or different to previous ones.
- Whether an emergency department visit or office follow-up was recommended.
- Whether a trial of NSAIDs or other analgesics was recommended.
- Whether the patient was told to call back with any change in symptoms.

REFERENCES

1. **Becker LA, Green LA, Beaufait D, et al.** Use of CT scans for the investigation of headache: a report from ASPN, Part 1. J Fam Pract. 1993;37:129-34.

2. **Rothrock SG, Buchanan C, Green SM, et al.** Cranial computed tomography in the emergency evaluation of adult patients without a recent history of head trauma: a prospective analysis. Acad Emerg Med. 1997;4:654-61.

3. **Ramirez-Lassepas M, Espinosa CE, Cicero JJ, et al.** Predictors of intracranial pathologic findings in patients who seek emergency care because of headache. Arch Neurol. 1997;54:1506-9.

4. **Attia J, Hatala R, Cook DJ, Wong JG.** Does this adult patient have acute meningitis? JAMA. 1999;282:175-81.

5. **Mikhail MG, Levitt MA, Christopher TA, Sutton MC.** Intracranial injury following minor head trauma. Am J Emerg Med. 1992;10:24-6.

6. **Mathew NT.** Differential diagnosis in headache-identifying migraine in primary care. Cephalalgia. 1998;18:22-32.

7. UpToDate 9.1, February 2001.

8. **Pryse-Phillips WE, Dodick DW, Edmeads JG, et al.** Guidelines for the diagnosis and management of migraine in clinical practice. Can Med Assoc. 1997;156:1273-87.

9. **Headache Classification Committee of the International Headache Society.** Classification and diagnostic criteria for headache disorders, cranial neuralgia, and facial pain. Cephalalgia. 1988;8:1-96.

10. **Lipton RB, Stewart WF, Ryan RE Jr, et al.** Efficacy and safety of acetaminophen, aspirin, and caffeine in alleviating migraine headache pain. Three double-blind, randomized, placebo controlled trials. Arch Neurol. 1998;55:210-7.

11. **Diamond S.** Ibuprofen versus aspirin and placebo in the treatment of muscle contraction headache. Headache. 1983;23:206-210.

12. **Miller DS, Talbot CA, Simpson W, Korey A.** A comparison of naproxen sodium, acetaminophen and placebo in the treatment of muscle contraction headache. Headache. 1987;27:392-6.

13. **Sargent JD, Peters K, Goldstein J, et al.** Naproxen sodium for muscle contraction headache treatment. Headache. 1988;28:180-2.

12

SKIN PROBLEMS

Michael S. Cohen, MD • David L. Stevens, MD

KEY POINTS

- Telephone diagnosis of skin problems requires taking a careful history and helping the patient describe the condition so the physician can form a "mental picture" of it.
- Rare but life-threatening skin-related conditions should be ruled out. These include toxic epidermal necrolysis (TEN), Stevens-Johnson syndrome (SJS), severe burns, and anaphylaxis.
- More common conditions such as cellulitis and zoster require disease specific treatment to prevent serious complications. When these conditions are considered likely, the patient should be seen in person within 24–72 hours depending on the suspected diagnosis and urgency of treatment.
- Many of the most common skin conditions, including most dermatitides, usually respond well to non-specific treatments such as topical steroids. Patients with these conditions should be seen in person if they do not respond to treatment.
- When available, over-the-counter preparations should be the first choice.

"Can you hold your rash up to the phone?" So goes the old joke. Yet in spite of the obvious challenge, dermatologists and primary care physicians frequently help patients with their skin problems over the telephone.

The diagnosis of a skin problem can often be a challenge even when the patient is in the room. Internists often do not recognize common dermatological conditions (1). Although the electronic transmittal of pictures of rashes promises to improve the accuracy of distance diagnosis in telemedicine, this is not yet a widespread practice and in its current form has limitations. One study showed that a dermatologist who examined a rash through a transmitted picture partially disagreed with a dermatologist who examined the patient's rash directly, in person, 3.5 times as often as he or she did when both examinations occurred in person (partial disagreement 21% vs. 6%) (2). Diagnosis without any visualization is especially challenging; nonetheless, careful history taking, including the use of the patient's own observations to develop a mental picture of the rash, can often yield sufficient information to permit proper triage and symptomatic treatment. The focus of this chapter is to help physicians use the patient's history and descriptions to rule out rare dermatological emergencies, determine which patients should be seen urgently (within 24-48 hours), and initiate symptomatic treatment in patients not requiring in-person evaluation. This chapter addresses the following central questions:

1. Does the patient require emergency evaluation for a possible life-threatening condition?
2. Does the patient require urgent evaluation (within 24-48 hours) for confirmation of diagnosis and disease-specific therapy?
3. When can treatment be recommended over the telephone?
4. How should skin problems be managed?

BACKGROUND

Epidemiology

Although the majority of skin conditions do not require hospitalization, there are a small number of life-threatening dermatoses (Table 12-1). Anaphylaxis, although not a purely dermatological problem, frequently presents with pruritus and hives and is included in this discussion for this reason. Burns are obviously much more common; the important issue is to

Table 12-1   Epidemiology of Life-Threatening Skin Conditions

Condition	Yearly Incidence	Mortality Rate (%)	Average Age	Typical Sex
Stevens-Johnson syndrome*	1.1 per million	34	25	Men 2:1
Toxic epidermal necrolysis[†]	0.93 per million	1	63	Women 2:1
Anaphylaxis[†]	200 per million	1	N/A	N/A

* Data from Reference 4.
[†] Data from References 5 and 6.

evaluate the severity (see below). Erythema multiforme majus (Stevens-Johnson syndrome [SJS]) and toxic epidermal necrolysis (TEN) are rare but the former carries significant risk of mortality.

The authors found no studies examining the prevalence of skin conditions of patients calling on the telephone, but clearly most skin conditions treated over the phone and in person are not emergencies. Conditions requiring urgent office evaluation are more common, such as cellulitis and zoster. The most common category of diagnosis is the dermatitides, which usually do not require emergency or even urgent evaluation.

Utility of Early Diagnosis

Early diagnosis and appropriate referral of the dermatological emergencies in Table 12-1 can be life-saving. Cases of TEN, SJS, severe burns, and anaphylaxis that are not recognized and treated early can result in severe infection, fluid and electrolyte derangements, and death.

Other conditions that are less of an emergency require specific therapy targeted to the etiology. Patients with conditions such as zoster and cellulitis seldom need emergency treatment (exceptions are listed below) but are at risk for further morbidity if disease-specific therapy such as antibiotics or antivirals is delayed. The risk of complications from these diseases, such as disseminated infection or post-herpetic neuralgia, increases when patients are not treated promptly (3). These patients should be seen in person to confirm the diagnosis and initiate therapy. Other conditions, such as severe poison ivy or oak, although not dangerous per se, may cause intolerable symptoms that respond only to oral corticosteroids. These patients should be seen promptly to confirm the diagnosis before initiating oral corticosteroids.

Fortunately, many patients will fall into the category of non-dangerous conditions that usually respond to non-specific treatments aimed at symptom relief. Usually these patients can be treated with over-the-counter therapies. The physician on the phone can provide significant relief of the patient's symptoms and reassurance that there is no cause for alarm, and save the patient lost time in needless visits to the emergency department.

INITIAL APPROACH TO THE TELEPHONE DIAGNOSIS

When confronted with a caller complaining of a rash or other skin problem, the information gathered falls into two broad categories: 1) information contributing to the physician's "mental picture" of the rash and 2) supportive information that will guide management (Table 12-2).

Creating a "Mental Picture" of the Condition

The main challenge in the telephone evaluation of a patient with a dermatological problem is to be able to visualize the patient's rash. Even patients with little medical experience can be very helpful in describing their skin

---

Table 12-2   Essential Information to Be Gathered from a Caller with a Skin Complaint

Information contributing to the physician's "mental picture"

- Lesion type
- Lesion shape
- Lesion color
- Lesion distribution
- Lesion arrangement

Supportive information that helps to guide management

- Pruritus or pain
- Chronicity: duration and seasonality
- Environmental exposure/travel
- Medications/cosmetics
- Past dermatological/medical history
- Possibility of pregnancy
- Constitutional symptoms, such as fever

findings. The challenge is to get a detailed description of the rash in the patient's own words. One obstacle is patients' use of non-medical terms such as "welt" or "pimple." A patient's understanding of these terms may differ from the physician's. In addition, a patient's description may be quite general, such as "I have a rash" or "I'm breaking out all over."

Questioning should begin with an open-ended invitation for the patient to describe the rash. For the reasons stated above, however, the patient's initial description may not help the physician visualize the lesions adequately. The key elements in the description are type, shape, distribution, arrangement, and color. By keeping these key elements in mind and asking a series of more direct questions using unambiguous terms, a recognizable pattern may emerge.

Key Elements of the Lesion Description

Lesion Type

The most common types of skin lesions are macules, papules, wheals, vesicles, and pustules. It is helpful to have the patient describe the lesions in as much detail as possible. When open-ended questions do not result in a clear mental image, the more specific questions in Table 12-3 may help the physician to discern the type of lesion.

A patient's estimation of the size of the lesions contributes to the physician's mental picture. The patient can be asked to compare the lesion to different-sized coins or a pea.

Lesion Distribution

The physician should ask the patient to describe where the skin lesions are located. When the patient says the rash is "all over," does this include

---

Table 12-3 Examples of Descriptive Questions About Skin Lesions

- Macule: *Are they flat?*
- Papule: *Are they bumps? Are they raised?*
- Wheal: *Are they like mosquito bites? Like hives?*
- Vesicle: *Are they blisters filled with water?*
- Pustule: *Do they have white heads?*

---

the arms, the legs, the back, and the chest? How about the scalp, the genitals, and face? If one can pinpoint where the rash is actually located, it is easier to assess for emergencies and manage non-emergencies. One should ask about eye involvement and oral mucosal involvement, which might suggest SJS. Palm and sole involvement may suggest SJS and TEN as well as syphilis. Eye/periorbital involvement often necessitates emergency evaluation.

Lesion Shape and Color

Shape and color are less helpful descriptions because most dermatological lesions tend to be red and round. Erythema chronicum migrans, indicative of Lyme disease, has a very characteristic ring surrounding uninvolved skin. Target lesions suggest erythema multiforme, which should prompt questioning about SJS and TEN (see below).

Lesion Arrangement

Arrangement of skin lesions sometimes suggests a diagnosis. Herpes zoster, for instance, will be dermatomal and unilateral, not crossing the midline. Poison ivy is characterized by the linearity of its skin lesions. Herpes simplex lesions are grouped. Patients may volunteer such information, but often more probing questions may be necessary, such as "Is it only on one side of your chest? None on the other side?" It is also important to remember that such characteristic arrangements can occur anywhere on the body, including the face, feet, and genitals.

Confirming the Physician's Mental Picture with the Patient

After exploring these key elements of the patient's rash, the physician should have a mental picture of the rash and should describe it back to the patient as a final check on the accuracy of the information acquired.

Supportive History Beyond the Lesion Description

Pruritus and Pain

An important element of the dermatological history is the patient's symptoms: pruritus (very common) or pain (rare). Most patients with a rash will admit to some itching. It is sometimes helpful to gauge the degree of pruritus from the patient. Generally, a good question to ask is "Is the itching keeping you awake at night?" or "Is the itching driving you crazy?" Severe pruritus would be more typical of urticaria, scabies, or bug bites. Pain, though less common, helps to key into a diagnosis such as herpes zoster; a good question might be "Does it hurt more than it itches?"

Chronicity: Duration and Seasonality

Generally, rashes with more rapid onset are more worrisome and more in need of urgent attention. Seasonal problems such sunburn, poison ivy, Lyme disease, and other conditions resulting from arthropod bites are obviously more common in warmer months because of variation in environmental exposures. An easily forgotten exception is winter vacations in warmer climates: a rash may not begin until the traveler returns home. Patients sometimes recall having a similar rash in a previous year in the same season, often related to environmental exposures. These patients may remember the diagnosis and treatment from their last physician encounter.

Medications and Cosmetics

Drug eruptions are quite common. A patient can develop an allergy to a medication at any time, but a medication started recently is more likely to be the source of the allergy. Any medication can cause a skin eruption, but antibiotics are generally thought to be most common. Rashes from antibiotics may take up to 14 days to develop, often after completion of the treatment, so it is important to detail all medications taken in the last month or two. A drug eruption can last for weeks after the drug was last taken.

All over-the-counter and topical therapies must be documented as well, including those taken to treat the condition. Many patients will have tried a topical medication that was in the medicine cabinet. The presence or absence of therapeutic effect may help with diagnosis. For example, a topical steroid that helps a rash fits with a dermatitis; a topical steroid that makes a rash worse suggests a fungal infection.

Cosmetic use should be reviewed. Even cosmetics that have been used for years may cause a skin reaction; especially if they are old, they may be colonized with bacteria.

Past Dermatological and Medical History

Patients may have an exacerbation of a pre-existing skin condition, such as psoriasis or eczema. These patients can often tell you what their skin problem is and how they have been successfully treated in the past. It is also important to know which treatments were unsuccessful. In addition, a patient with a pre-existing skin condition may not recognize the new problem as being the same problem, so it is important to ask if a patient has a history of skin problems. A patient with a long history of psoriasis in certain locations or configurations may not recognize psoriasis that has presented in a different way.

Chronic medical conditions may also affect management, especially if there is a possibility of impaired immunity or wound healing. Diabetes mellitus, chronic oral corticosteroid use, and impaired immunity from AIDS or chemotherapy all dictate a greater degree of caution and a lower threshold for in-person evaluation.

Risk of Pregnancy

A number of dermatological conditions may present significant risk to the fetus (see below).

Constitutional Symptoms

Constitutional symptoms may provide a clue to the diagnosis. For example, viral symptoms, such as fever, malaise, nausea, and headache, are more likely to occur with a viral exanthem.

Fever and a rash of unclear etiology has a large differential diagnosis, including bacterial infections. These patients should generally be seen within 24 hours to evaluate the need for antibiotics.

DOES THE PATIENT REQUIRE EMERGENCY EVALUATION FOR A POSSIBLE LIFE-THREATENING CONDITION?

As stated above, dermatological conditions can be divided into those requiring emergency evaluation and treatment, those necessitating urgent evaluation, and those that are appropriate to diagnose and treat over the phone. Emergency conditions include anaphylaxis, severe burns, and erythema multiforme majus.

Could the Patient Have Anaphylaxis?

Any patient with possible urticaria should also be asked about systemic symptoms. Common causes are medications (most typically antibiotics, NSAIDs, and opioids [6]), foods, and bee stings. Penicillin is the leading cause of fatal anaphylaxis (6). Patients with possible urticaria should be asked if they have shortness of breath, tightness in the throat, difficulty in swallowing, facial swelling, or new gastrointestinal symptoms. The physician should note any unusual sounds with breathing (stridor or wheezing). Any such findings should prompt an emergency department visit for evaluation and possible treatment with epinephrine and antihistamines. Because of the rapid course of anaphylaxis, treatment should be initiated immediately, even before getting to the emergency department (see section on management). Patients with hives without additional symptoms should be advised to watch for such symptoms and call back immediately or go to the emergency department should they occur.

Does the Patient Have a Severe Burn?

Most burns are minor; only 5% require hospitalization (7). A small burn that is erythematous may need no more than application of a topical antibiotic ointment or even just a skin lubricant. On the other hand, a burn on the face, genitals, or hands, or a burn covering a significant surface area needs prompt evaluation and may require the services of a burn center.

The American Burn Association guidelines recommend hospitalization for moderate or severe burns (see Table 12-4). These patients should be referred immediately to an emergency department, preferably a burn center. Minor burns that go beyond the most superficial layer of skin should also be seen to evaluate the need for debridement to reduce the risk of infection. This evaluation should be performed as soon as possible but can be accomplished in an outpatient setting,

Table 12-4   American Burn Association Criteria for Hospitalization in Adults

- Greater than 10% total body surface area
- Greater than 2% full-thickness burn
- High-voltage injury
- Suspected inhalation injury
- Circumferential burn
- Concomitant medical problem predisposing to infection
- Any significant burn to face, eyes, ears, genitalia, or joints
- Significant associated injuries (other major trauma)

Data from Reference 7.

provided the physician has the proper experience and equipment. Table 12-5 presents a rough guide for gauging the depth of burns. Superficial burns usually respond to topical measures (see section on management below).

The total body surface area can be estimated using the rule that the patient's palm represents approximately 0.8% of the total body surface area. Clearly, exact measurements are not feasible over the phone. Nonetheless, this should allow the physician to be more confident advising the caller with a small superficial burn on home management.

Could the Patient Have Stevens-Johnson Syndrome or Toxic Epidermal Necrolysis?

Although these conditions are rare, they carry significant risk of morbidity and mortality. Although no studies are available to suggest a method for ruling these conditions out over the phone, certain general diagnostic principles are well accepted. Generally, both conditions are associated with drugs such as antibiotics, NSAIDs (4), and anticonvulsants (8). Any patient who reports a rash with features of possible erythema multiforme minor (urticaria, blisters, target lesions, lesions on the palms or soles) needs to be questioned about the presence of blisters or erosions on the oral mucosa or ocular mucosa, characteristic of SJS.

Table 12-5   Classification of Burns Based on Depth

Classification	Cause	Characteristics	
		Appearance	*Sensation*
Superficial burn	Ultraviolet light, very short flash (flame exposure)	Dry and red; blanches with pressure	Painful
Superficial partial thickness burn	Scald (spill or splash), short flash	Blisters; moist, red and weeping; blanches with pressure	Painful to air and temperature
Deep partial thickness burn-	Scald (spill), flame, oil, grease	Blisters (easily unroofed); wet or waxy dry; variable color (patchy to cheesy white to red); does not blanch with pressure	Perception of pressure only
Full-thickness burn	Scald (immersion, flame, steam, oil, grease, chemical, high-voltage electricity)	Waxy white to leathery gray to charred and black; dry and inelastic; does not blanch with pressure	Deep pressure only

Modified from Morgan ED, et al. Ambulatory management of burns. Am Fam Phys. 2000;62:2015-26.

TEN should be considered in a patient complaining of skin sloughing, typically in sheets, no matter how small a surface area. In both TEN and SJS, prompt diagnosis is vital because either condition may progress in a matter of hours. These patients require inpatient monitoring, frequently in an ICU or a burn-unit setting. Intensive topical treatments are necessary as well as intravenous fluids and possibly intravenous antibiotics and steroids.

DOES THE PATIENT REQUIRE URGENT EVALUATION FOR CONFIRMATION OF DIAGNOSIS AND DISEASE-SPECIFIC THERAPY?

If a patient does not require immediate emergency department or office evaluation, he or she may still benefit from in-person evaluation within 24-48 hours. These patients have conditions that carry a risk of

---

Table 12-6   Conditions That Should Be Seen Within 24-48 Hours

- Cellulitis (see within 24 hours)
- Herpes zoster/shingles (see within 72 hours of onset)
- Lyme disease
- Severe poison ivy or oak
- Unexplained rash with pregnancy
- Inflamed or infected cysts, boils, or furuncles
- Fever and unexplained rash
- Moles that have changed (see within a week)

---

complications if untreated and/or require medications that should not be prescribed without in-person confirmation of diagnosis, such as oral corticosteroids, antibiotics, or antivirals (see Table 12-6). In addition, in-person evaluation allows the physician to monitor the response to treatment more effectively. An exception to the recommendations below is periorbital involvement of the rash; in most cases this should be seen immediately.

It is beyond the scope of this book to list all the conditions requiring in-person evaluation, but the discussion below includes the most common.

Could the Patient Have Cellulitis?

Cellulitis usually requires oral or intravenous antibiotics. A patient who reports symptoms of limb redness, swelling, warmth and pain, or tenderness should be presumed to have cellulitis and requires antibiotics. Intravenous antibiotics may be necessary, especially if the patient has concomitant fever, chills, or sweats or if the patient is immunocompromised or diabetic. Physicians should not prescribe oral antibiotics on the phone even for probable uncomplicated cellulitis because only in-person evaluation can confirm the diagnosis and rule out other "look alike" conditions such as deep venous thrombosis or zoster (see below). Patients with suspected cellulitis should be seen within 24 hours.

Could the Patient Have Herpes Zoster?

Herpes zoster is usually diagnosed by dermatomal involvement of erythematous plaques and clustered vesicles and the presence of pain. Diagnosis may be more challenging with involvement of the face, genitals, or lower extremity. Prompt treatment with oral antiviral agents within 72 hours can reduce acute pain, viral shedding, and post-herpetic neuralgia (3). If the diagnosis seems likely based on the history but in-person evaluation is not possible within 72 hours of onset, oral antivirals may be started, with office follow-up within 1-2 days to confirm the diagnosis (see section on management).

Could the Patient Have Dermatitis Caused by Poison Ivy or Poison Oak?

Poison ivy and poison oak usually require super-potent topical steroids and sometimes systemic corticosteroids. A history of recent gardening or hiking suggests the diagnosis. There is a characteristic eruption of linear urticarial plaques and vesicles, usually with intense pruritus. A large involved surface area or facial or genital involvement may necessitate systemic steroids. Although oral corticosteroids can be started over the phone, the diagnosis should be confirmed in person to avoid unnecessary treatment because the appropriate length of treatment is usually 2 weeks.

Is the Patient Pregnant?

Serious dermatological conditions in pregnant women, such as herpes gestationis and maternal varicella, can cause fetal death or congenital abnormalities (8) and require rapid initiation of medication. Two of the more serious dermatological conditions in pregnancy are discussed below.

• *Herpes gestationis* is an extremely pruritic and rare papulovesicular eruption that is commonly located on the abdomen, usually during the second trimester of pregnancy. Systemic steroids may be necessary. Despite its name, it is unrelated to any herpes virus.

- *Maternal varicella* can be dangerous to the fetus and may require systemic acyclovir. Diagnosis should be made promptly. A reliable history of previous varicella is helpful in ruling out the diagnosis.

Given the high risk and the limitations of telephone diagnosis, physicians should consider emergency department or office evaluation within 24 hours for any pregnant patient with a rash of uncertain origin.

Does the Patient Have Inflamed or Infected Cysts, Boils, or Furuncles?

These may require surgical incision and drainage of the lesion. Usually warm compresses can be initiated over the phone, but incision and drainage is generally eventually necessary. Facial involvement requires more rapid evaluation.

Does the Patient Have Rash Accompanied by Fever?

Fever and rash carry a long differential diagnosis, including bacterial, viral, and rickettsial infections. The diagnosis is often challenging, even in person. Because disease-specific medication is often necessary, patients with fever and unexplained rash should be seen within 24 hours.

Could the Patient Have Lyme Disease?

Patients may call with concerns specifically about Lyme disease, whether or not they actually have the typical rash. Prescription of antibiotics over the phone may be reasonable for some of these callers, but most patients with tick bites are at low risk. A study in Westchester County, New York, one of the areas where the incidence of Lyme disease is highest, found that Lyme disease developed in only 3.2% of people with deer tick bites (11,12).

The risk of Lyme disease after a deer tick bite can be better assessed if the duration of the bite can be estimated. The risk is higher if the tick appears engorged (9.9%) or if it has been feeding for 72 hours or longer (25%). If the tick is flat or is removed within 72 hours, the risk is 0% (11). Furthermore, tick bites outside of the northeast United States carry a much lower risk for Lyme disease.

A challenge to the physician on the telephone is to assess whether the tick is engorged or even if the insect was indeed a deer tick. However, if the patient is confident the tick was removed within the 72 hours, antibiotics are unnecessary. Furthermore, antibiotics can safely be deferred until a rash develops: there seems to be no greater risk of long-term complications (11).

A patient with erythema migrans (erythema with central clearing) who was bitten by a tick in an endemic area may have Lyme disease. Patients typically also have symptoms suggestive of a viral syndrome (9). Physicians and patients have been very aggressive about early treatment of Lyme disease based on fear of long-term complications. Although it is improbable that waiting a few days for an office visit will significantly increase this risk, physicians may decide to initiate treatment in specific individuals, such as patients calling from remote areas who cannot come in within a few days. In these cases, antibiotics can be started, but the patient should still be encouraged to come in for an in-person evaluation of the rash as soon as possible.

Does the Patient Have a Mole That Has Changed?

Patients will often call about a mole that has changed. This may be a seborrheic keratosis or skin tag that has become irritated. The lesion may recently have bled, become larger, bumpier, or darker. In order to rule out a malignant melanoma, the patient should be seen in the office promptly, but this does not require emergency intervention and can certainly wait a few days.

WHEN CAN TREATMENT BE RECOMMENDED OVER THE TELEPHONE?

There are numerous dermatological conditions that can be treated simply over the phone. A good general approach when these conditions are suspected is to try home management and have the patient seen in the office within a few days if the condition is not improving. These conditions are commonly encountered, have low risk of progressing to more severe conditions, and respond well to the non-specific treatments discussed later in this chapter. This category includes such common acute conditions as hives, bug bites, contact dermatitis, and pityriasis rosea. Symptomatic treatments are described below.

HOW SHOULD SKIN PROBLEMS BE MANAGED?

General Principles of Management

A variety of medications and other supportive measures are available for symptom relief. Whenever disease-specific treatment is warranted, the physician should try to confirm diagnoses with in-person evaluation whenever possible.

Over-the-counter preparations should be the first choice because they are generally safe, inexpensive, and easy to use. Many dermatological conditions can be treated temporarily with over-the-counter remedies until the patient can be evaluated in the office.

Combination over-the-counter skin therapies may contain unnecessary active ingredients that may cause topical sensitization. Patients should be advised to purchase medications containing only the recommended active ingredient.

Disease-Specific Treatments

Anaphylaxis

When anaphylaxis is suspected, the patient should immediately take an antihistamine, if one is available, such as diphenhydramine 50 mg, and proceed to the emergency department. When there is any evidence of cardiovascular or respiratory distress, 911 should be called, either by a family member or the physician. Causative agents, such as a bee stinger, should be removed if possible to prevent further introduction of toxin. Patients with a history of anaphylaxis who have epinephrine pens at home should use them.

Burns

Superficial burns can be managed with topical lubricants such as aloe vera, analgesics such as NSAIDs, and antipruritics such as diphenhydramine (7). Some deeper burns (superficial partial thickness, see Table 12-5) can be debrided and dressed as an outpatient. Burns meeting the criteria in Table 12-4 should be referred to a burn center.

Bacterial Infection: Topical and Oral Antibiotics

Topical Antibiotics
Triple antibiotic ointment is usually an effective topical treatment for many uncomplicated superficial wounds.

Oral Antibiotics/Antivirals

LYME DISEASE
Patients living in endemic areas with a typical rash and a tick bite may be started on antibiotics. Doxycycline (100 mg bid) or amoxicillin (for pregnant women) (500 mg tid) for 14-21 days is usually adequate. Diagnosis should be confirmed in person whenever possible. Patients with tick bites but no rash generally do not need prophylaxis, especially if the tick is removed within 72 hours. If the physician chooses prophylaxis, a single dose of doxycycline 200 mg is an effective option for non-pregnant adults within 72 hours of the bite (11).

HERPES SIMPLEX AND ZOSTER
Herpes simplex, most commonly a recurrent infection, presents with painful blisters on a well-demarcated erythematous base. Most commonly, the lip is involved. Outbreaks may be triggered by sunburn or acute illness. The patient will often have a history of previous "fever blisters" or "cold sores." Given a typical presentation in a patient with a history of the same type of lesion, the diagnosis is often clear, and prescribing an antiviral over the phone is reasonable. When started within 72 hours of the onset of the eruption, oral antiviral medications may decrease the duration of lesions, fever, and viral shedding (10). Alternatively, a topical anaesthetic such as benzocaine or viscous lidocaine can be helpful in reducing discomfort.

Oral acyclovir, famciclovir, and valacyclovir are all effective for both zoster and recurrent herpes simplex eruptions. Dose reductions are necessary in renal insufficiency. Table 12-7 lists the doses for these medications.

Non-Specific Treatments

Topical Steroids

Topical hydrocortisone 1% is available over the counter and may be effective for many inflammatory skin disorders such as contact dermatitis and eczema. Hydrocortisone serves as a good first-line treatment until the patient can be seen and switched to a more potent topical steroid if necessary. These preparations should be avoided, however, if the condition is suggestive of fungal infection, such as long-standing erythema in an intertriginous area.

Oral Antihistamines

Oral antihistamines (both first and second generation) are effective and safe treatment for urticaria and other pruritic dermatoses. Diphenhydramine is available over the counter 25-50 mg up to 4 times a day. Histamine-2 antagonists, such as ranitidine 150 mg bid, can be useful in conjunction.

Topical Anti-Pruritics

Topical therapies containing menthol, such as Sarna or Pramegel, are very effective anti-pruritic therapies.

Cold-Milk Compresses

Cold-milk compresses are helpful in most pruritic conditions, often provide nearly immediate relief, and have no restrictions in terms of how often they are used. They involve no risk to the patient and no trip to the drugstore in the middle of the night. The patient can soak a clean towel in whole milk with ice cubes and apply to the affected areas as often as needed.

Table 12-7   Doses for Treatment of Zoster and HSV

	HSV Recurrence	*Herpes Zoster (Shingles)*
Acyclovir	200 mg 5×/d for 5 days	800 mg 5×/d for 7–10 days
Famciclovir	125 mg bid for 5 days	500 mg tid for 7 days
Valacyclovir	500 mg bid for 5 days	1 g tid for 7 days

From Drug Information Handbook. Lacy C. Lexi-Company 2000; with permission.

WHAT TO TELL THE PATIENT

Patient Requires In-Person Evaluation

- *Explain to the patient why he or she needs to be seen:* "From your description, it sounds like you may have something serious such as [SJS/TEN, cellulites, zoster, etc.]. Examining you in person would help me make sure."
- *Explain the risks of the condition:* "If this is the condition I'm concerned about, it may progress to [severe infection, etc.]."
- *Describe the treatments that are available:* "Treatment such as [indicate the treatment] may prevent this from progressing and make you feel better."

Patient Does Not Require In-Person Evaluation

- *Low-risk condition:* "From your description, it doesn't sound like this is anything dangerous. Without examining you in person, it's hard to be sure, but you don't describe any of the features of a more dangerous condition."
- *Home management:* "I recommend that you try some things at home to see if it gets better, such as [indicate home treatment]."
- *What to watch for/when to call back:* "I also recommend you keep an eye on it. If the rash changes, or if it doesn't get better over the next day or two, call back and we'll arrange an office visit."

PATIENT WHO CALLS WITH SKIN PROBLEM

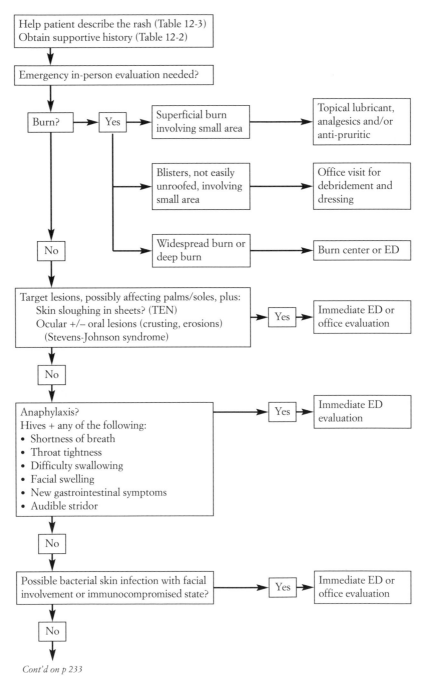

Cont'd on p 233

Cont'd from p 232

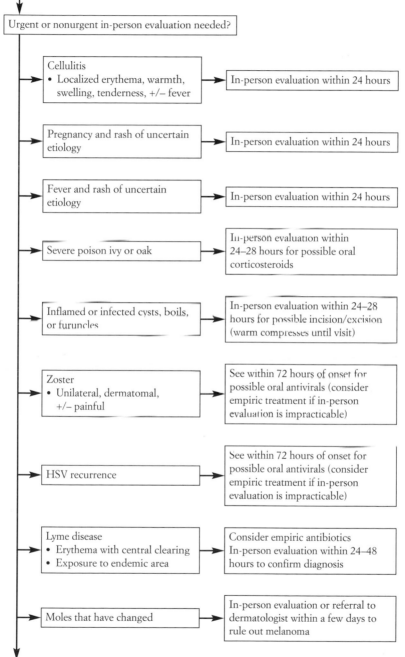

Urgent or nonurgent in-person evaluation needed?

Cellulitis
• Localized erythema, warmth, swelling, tenderness, +/– fever
→ In-person evaluation within 24 hours

Pregnancy and rash of uncertain etiology
→ In-person evaluation within 24 hours

Fever and rash of uncertain etiology
→ In-person evaluation within 24 hours

Severe poison ivy or oak
→ In-person evaluation within 24–28 hours for possible oral corticosteroids

Inflamed or infected cysts, boils, or furuncles
→ In-person evaluation within 24–28 hours for possible incision/excision (warm compresses until visit)

Zoster
• Unilateral, dermatomal, +/– painful
→ See within 72 hours of onset for possible oral antivirals (consider empiric treatment if in-person evaluation is impracticable)

HSV recurrence
→ See within 72 hours of onset for possible oral antivirals (consider empiric treatment if in-person evaluation is impracticable)

Lyme disease
• Erythema with central clearing
• Exposure to endemic area
→ Consider empiric antibiotics In-person evaluation within 24–48 hours to confirm diagnosis

Moles that have changed
→ In-person evaluation or referral to dermatologist within a few days to rule out melanoma

Cont'd on page 234

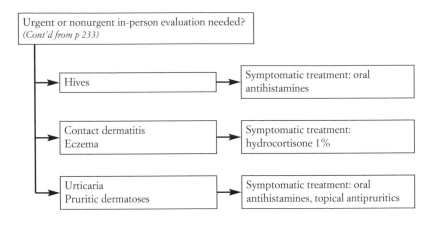

Urgent or nonurgent in-person evaluation needed?
(Cont'd from p 233)

Hives → Symptomatic treatment: oral antihistamines

Contact dermatitis
Eczema → Symptomatic treatment: hydrocortisone 1%

Urticaria
Pruritic dermatoses → Symptomatic treatment: oral antihistamines, topical antipruritics

WHAT TO DOCUMENT

- *Description of symptoms:* quality of rash, location, associated symptoms, and chronology.
- *Medical history and medications:* whether patient is pregnant, chronic illnesses, past dermatological history, and prescription and non-prescription medications, including topicals.
- *What you told the patient:* low vs. high probability of dangerous etiology; instructions on if, where, and when the patient should be evaluated in person; home management; and when to call back or come to the office.
- *Plan:* whether patient agreed with your recommendations and, if not, what patient intended to do.

REFERENCES

1. **Kirsner RS.** Lack of correlation between internists' ability in dermatology and their patterns of treating patients with skin disease. Arch Dermatol. 1996;132: 1043-6.
2. **Lesher JL.** Telemedicine evaluation of cutaneous diseases: a blinded comparative study. J Am Acad Dermatol. 1998;38:27-31.
3. **Crovo D, Bajwa ZH, Warfield CA.** Pain associated with herpes zoster infection. Up To Date; 2000.
4. **Schopf E.** Toxic epidermal necrolysis and Stevens-Johnson syndrome. An epidemiologic study from West Germany. Arch Dermatol. 1991;127:839-42.

5. **Yocum MW.** Epidemiology of anaphylaxis in Olmsted County: a population-based study. J Allergy Clin Immunol. 1999;104:452-6.
6. **Neugut AI, Ghatak AT, Miller RL.** Anaphylaxis in the United States. Arch Intern Med. 2001;161:15-21.
7. **Morgan ED, Bledsoe SC, Barker J.** Ambulatory management of burns. Am Fam Phys. 2000;62:2015-26.
8. **Fitzpatrick TB, Eisen AZ, Wolff K, et al.** Dermatology in General Medicine. New York: McGraw-Hill; 1999.
9. **Sigal LH.** Treatment of Lyme disease. Up To Date; 2000.
10. **Klein RS.** Treatment and prevention of herpes simplex virus type I infection. Up To Date; 2000.
11. **Shapiro ED.** Doxycycline for tick bites—not for everyone. N Engl J Med. 2001;345:133-4.
12. **Nadelman RB, Nowakowski J, Fish D, et al.** Prophylaxis with single-dose doxycycline for the prevention of Lyme disease after an ixodes scapularis tick bite. N Engl J Med. 2001;345:79-84.

13

EMERGENCY CONTRACEPTION
AND OTHER POST-COITAL
CONSIDERATIONS

Cary P. Gross, MD

KEY POINTS

Emergency Contraception

- Many women, and many physicians, are not aware of emergency contraception.
- Oral contraceptives (OCs) can reduce the risk of pregnancy by 60% to 70% when the first dose is taken within 72 hours of unprotected intercourse.
- The intrauterine device (IUD) can reduce the risk of pregnancy by 98% if inserted up to 5 days after intercourse.
- Patients should be counseled about the importance of resuming safer sex practices, with effective contraception and attention to preventing sexually transmitted diseases.

HIV

- HIV prophylaxis can be highly effective but should be used primarily in patients who were exposed to people with known HIV or who have a high likelihood of infection.
- If there is a risk of exposure to HIV, the patient should seek medical attention in an emergency department or a doctor's office as soon as possible.

Continued

- The partner should also be encouraged to be tested, if HIV status is not known.
- If prophylaxis is indicated, it should be administered within hours of the exposure.

Hepatitis B
- The regimen of hepatitis B immunoglobulin (HBIG) and hepatitis B virus (HBV) vaccine is thought to be effective for as long as 14 days after exposure.
- The patient should be instructed to see his or her physician the next business day and to only present to the emergency department if the 14-day deadline occurs before then.

M any questions about the implications of unprotected sex or broken condoms are likely to arise on evenings or weekends. This chapter will mainly focus on emergency contraception through the use of oral contraceptives, which were approved by the FDA for emergency post-coital contraception (EPC) in 1997. Because EPC should be initiated within 72 hours of the event, patients may not have the luxury of waiting for an office appointment. Likewise, post-exposure prophylaxis for HIV and hepatitis B is a subject the telephone medicine practitioner may have to address. Physicians should be comfortable discussing the options and initiating treatment over the telephone. This chapter focuses on the following questions:

- Which patients should receive emergency contraception?
- What alternatives are currently available for emergency contraception?
- Which patients should receive post-exposure prophylaxis for HIV?
- Which patients should receive post-exposure prophylaxis for hepatitis B?

This chapter will also describe how to counsel a woman who is concerned about a recent exposure to HIV or hepatitis B virus.

EMERGENCY POST-COITAL CONTRACEPTION

Epidemiology

It has been estimated that the widespread use of EPC could prevent up to 3 million unwanted pregnancies annually in the United States (1). Although oral contraceptives have been used for EPC for more than 20 years, the FDA did not approve the marketing of hormonal regimens specifically for EPC until 1997. Many physicians and patients are still unfamiliar with EPC (2).

What Is the Risk of Pregnancy When Contraception Is Not Used?

When women are deciding whether to use EPC, it may be useful to consider the risk of pregnancy. Untimed intercourse averaging once per week for one menstrual cycle results in conception in about 15% of women (3). The fertile period lasts for approximately 6 days and ends on the day of ovulation, which usually occurs 14 days before the onset of menstrual flow (3,4). The probability of conception is highest (approximately 30%) when intercourse occurs on the day of ovulation or during the previous 2 days.

Does the Patient Need Emergency Contraception?

Eligible women include those who have had unprotected sex in the previous 72 hours and are not currently pregnant. EPC may also provide an important safety net when other forms of contraception fail: when a condom breaks, a diaphragm slips out of place, or a contraceptive injection is more than 2 weeks late. Because so few women know about EPC, the focus of the call may be different, such as concern about possible exposure to STDs or HIV (see below). EPC should always be discussed with patients concerned about recent unprotected sex, and patients should be provided with enough information to make an informed decision.

EPC is usually not indicated for women who use OCs regularly but who have missed 2 or 3 days (2). Such women should instead be instructed to take the most recent missed pill, discard earlier missed pills, and use additional contraceptive precautions (i.e., barrier method) until the next menstrual period. If there were fewer than seven (active) pills left in the packet, the patient should begin the next cycle without the usual 7-day waiting period (2).

What Are the Options for Emergency Post-Coital Contraception?

Combined Estrogen-Progesterone Formulations

The combined estrogen-progesterone regimens that are available for EPC are listed in Table 13-1.

Is This Contraception or an Abortifacient?
Misconceptions about the mechanism of action may needlessly decrease EPC use by physicians and patients who are concerned about the distinction between abortion and contraception. The American College of Obstetrics and Gynecology defines pregnancy as beginning with implantation.

Table 13-1   Hormonal Regimens for Emergency Contraception

Generic Name	Brand Name	Color of Tablet	Tablets per Dose
Combined estrogen-progestin regimen			
Norgestrel 0.50 mg + ethinyl estradiol 25 µg	Preven Kit*	Light blue	2
Norgestrel 0.50 mg + ethinyl estradiol 50 µg	Ovral	White	2
Norgestrel 0.30 mg + ethinyl estradiol 30 µg	Lo/Ovral	White	4
Levonorgestrel 0.15 mg + ethinyl estradiol 30 µg	Levlen	Light orange	4
	Nordette	Light orange	4
	Levora	White	4
Levonorgestrel 0.10 mg + ethinyl estradiol 20 µg	Alesse	Pink	5
Levonorgestrel 0.125 mg + ethinyl estradiol 30 µg	Tri-Levlen	Yellow	4
	Triphasil	Yellow	4
	Trivora	Pink	4
Progestin-only regimen			
Norgestrel 0.75	Plan B	White	1

* Includes a home pregnancy test.

Although the exact mode of action for high doses of estrogen and proges-terone in preventing pregnancy is unclear, likely mechanisms include inhi-bition of ovulation and interference with implantation (2,5). Because emergency contraception prevents ovulation and implantation, it should be considered a form of contraception rather than an abortifacient (2,5). Some women, however, may consider any agent that functions after con-ception to be an abortifacient. The physician should be specific in stating that EPC does not necessarily prevent conception.

What Is the Efficacy of EPC?

A recent meta-analysis (of 7 studies that included more than 2800 patients) concluded that the combined estrogen-progestin EPC regimen reduced the expected number of pregnancies by approximately 75% (95% confidence interval: 66% to 82%) (6). Previous studies have suggested that initiating therapy earlier in the 72-hour interval may lead to increased efficacy. There-fore, women should be counseled to initiate therapy as early as possible.

Are There Risks If the Patient Is Already Pregnant?

EPC should not be used if pregnancy is detected. It is not believed to be harmful, but it will not be effective (22). Although teratogenicity of the EPC regimen has not been studied specifically, studies of combined oral contraceptives have failed to demonstrate teratogenicity (7,8), and the FDA has removed warnings about adverse effects of oral contraceptives on the fetus from the package insert.

Should the Patient Take a Pregnancy Test Before Use?

Some sources recommended that women take a home pregnancy test before initiating EPC. The Preven Emergency Contraception Kit, which is specifically packaged and marketed for use as EPC, includes a home pregnancy test. However, a pregnancy test need not be done in women who have had a normal menstrual period in the past 4 weeks (9).

What Are Common Side Effects and Contraindications of Oral EPC?

The major side effects associated with the use of combined oral contra-ceptive regimen are nausea (50%) and vomiting (approximately 20%) (2). Because of the increased risk of thromboembolism in women taking oral contraceptives on a regular basis, there is a theoretical concern over their use as EPCs. Therefore, relative contraindications to the combined EPC are pre-existing venous thromboembolic disease and active migraine with marked neurologic symptoms (10). In these instances, it is advisable to recommend use of the progestin-only regimen or an IUD (see below). All EPC patients should be educated about the warning signs of less common but important conditions associated with pregnancy (e.g., ectopic pregnancy) and use of estrogen (e.g., thromboembolism).

Progestin-Only Regimen (Formerly Known as the "Mini-Pill")

How Does the Progestin-Only Regimen Differ from Estrogen-Progestin Formulations?
Levonorgestrel, given alone in two separate doses of 0.75 mg each, is an attractive alternative to the combined regimen. A large multicenter randomized trial study by the World Health Organization recently concluded that the progestin-only regimen may be more effective (11). Additionally, patients receiving the progestin-only regimen were less likely to complain of nausea (23.1% vs. 50.5%) and vomiting (5.6% vs. 18.8%) than those receiving the estrogen-progestin regimen. A progestin-only regimen was recently approved by the FDA for marketing using the name "Plan B" (12). The higher efficacy and lower incidence of side effects may justify choosing the progestin-only regimen. For women who cannot take estrogen, the progestin-only regimen is an excellent alternative.

Intrauterine Device

How Is an Intrauterine Device Used for Emergency Contraception?
Although placement of a copper-T intrauterine device (IUD) cannot be prescribed over the telephone, clinicians should be familiar with its use as an emergency contraceptive device. The IUD reduces the risk of pregnancy after unprotected intercourse by more than 99% (2). Additionally, a copper-T IUD can be left in place to provide continuous effective contraception for up to 10 years.

What Are Its Advantages over the Oral EPC Regimens?
Unlike the oral regimens, the IUD can be inserted up to 5 days after unprotected intercourse to prevent pregnancy. This provides a useful

window of time for women who are more than 72 hours past their most recent intercourse and wish to receive emergency contraception.

What Are the Contraindications to IUD Use as Emergency Contraception?
IUDs are not ideal for all women. Women at increased risk of sexually transmitted infections may not be good candidates because insertion of the IUD is associated with an increased risk of pelvic inflammatory disease and infertility. Insertion of IUDs can be difficult in young, nulliparous women, and the cost of IUD insertion is higher than an oral contraceptive regimen. Confirmed or suspected pregnancy is a contraindication for insertion of an IUD.

What If More Than 72 Hours Have Passed and the IUD Is Not an Option?
There are sparse data about the efficacy of oral contraceptives for EPC in this setting. There may be potential benefit even up to 1 week after intercourse (5).

CAN RU-486 (MIFEPRISTONE) BE USED FOR EMERGENCY POST-COITAL CONTRACEPTION?

Although the antiprogestin RU-486 was recently approved by the FDA for medical abortion, it has not been approved for EPC. However, some preliminary studies have suggested that RU-486 is an effective and safe alternative for EPC (2). Further studies are warranted to compare RU-486 with existing treatment options.

WHICH EMERGENCY POST-COITAL CONTRACEPTION REGIMENS MAY BE PRESCRIBED OVER THE TELEPHONE?

There are many oral contraceptive formulations on the market (Table 13-1). Patients should be instructed to take the tablets with the correct color because only those contain the active hormones (13). The U.S. Food and Drug Administration (FDA) has approved both Preven and Plan B (progestin) for use and marketing for EPC. The remaining combined estrogen-progestin regimens listed in Table 13-1, currently marketed as birth control pills, have been declared safe and effective by the FDA for use as EPC.

WHAT TO TELL THE PATIENT

It is important to make certain that the patient does not want to become pregnant. Ask her if she is in a place where she feels comfortable talking and asking questions because she may not be able to talk freely at work or at home. After explaining the mechanism of the contraceptive (e.g., that it prevents implantation and not conception), explore her understanding of the process and ask if she feels comfortable proceeding. Explain that it is difficult to estimate the risk of pregnancy because predicting the day of ovulation is often unreliable (14). She should also be encouraged to discuss the decision with her partner, if appropriate. Document her concerns as well as what she was told about how EPC works.

The patient should be given the following instructions for taking the oral EPC regimen:

- *"Nausea and vomiting are common side effects.* You may experience nausea and vomiting; these symptoms usually resolve within 24 hours. An anti-nausea medication such as meclizine hydrochloride (Dramamine II, Bonine: 50 mg) 1 hour before the first dose may prevent these symptoms. If you vomit within 2 hours of taking EPC, call your clinician; you may need to have an additional dose prescribed."
- *"Take the first dose within 72 hours of unprotected sex, and take the second dose 12 hours after the first dose.* Initiating therapy as early as is possible will increase the likelihood of preventing pregnancy. However, remember that you will need a second dose 12 hours later. For example, taking the first dose at 4:00 p.m. would necessitate awakening at 4:00 a.m. for the second dose. It may be easier to wait until later in the evening before taking the first dose. Do not take extra EPCs. They are unlikely to decrease the risk of pregnancy, and they will increase the risk of nausea."
- *"Your next menstrual period may start a few days earlier or later than usual.* Emergency contraception is not 100% effective. If

PATIENT WHO CALLS AFTER UNPROTECTED INTERCOURSE

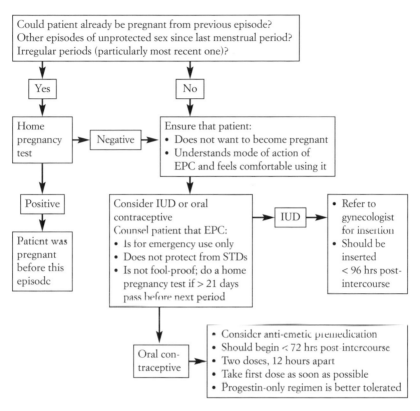

more than 21 days elapse before the next menstrual period begins, call your clinician for an exam and a pregnancy test."

- *"Watch for 'danger signs' during the next few weeks.* Contact your clinician immediately if you experience severe abdominal pain, severe leg pain, severe chest pain, cough, shortness of breath, severe headaches, dizziness, weakness, numbness, difficulty speaking, or visual changes."

- *"Emergency contraception is for emergency use only.* You may be fertile during the first several days after taking EPC. Therefore, you should begin using contraception on a regular basis, beginning with the next time you have intercourse. Emergency contraception should not be used as a regular contraceptive because other methods of contraception are more effective, better tolerated, and, in the case of condoms, provide protection against sexually transmitted diseases."

- *"Emergency contraception does not protect against or treat sexually transmitted diseases.* If you are at risk for or think you may have symptoms of a sexually transmitted disease, make an appointment to see your physician. If you think that you may have been exposed to hepatitis B and HIV, you may benefit from immediate evaluation and treatment."

- *"Insertion of an IUD is another option for emergency contraception.* The IUD is the most effective means of post-coital contraception and also provides contraception on a regular basis after insertion, unlike hormonal EPCs. If you are interested in IUD insertion, you must see your gynecologist."

WHAT TO DOCUMENT

- When last intercourse occurred
- Last menstrual period and whether periods were irregular
- Whether a home pregnancy test was done
- Whether patient definitely does not want to become pregnant
- Patient understanding of EPC and comfort with its use
- Whether patient preferred EPC with oral contraceptives or IUD
- Whether counseling on EPC (what it will and will not protect from) was done
- What was prescribed and whether instructions and side effects were understood by the patient
- Follow-up plan

HIV PROPHYLAXIS

Epidemiology

As of June 1999, more than 279,000 people have been diagnosed with AIDS in the United States (15). Approximately 10% of these infections were acquired through heterosexual contact (15). For women, heterosexual contact remains the most common route of exposure, with almost twice the number of cases as injection drug use (15).

How Can You Evaluate the Risk of Acquiring HIV After Sexual Contact?

The likelihood of contracting HIV from a specific sexual contact depends on a number of factors, including characteristics of the contact, the type of sex, and characteristics of the exposed person.

If the contact is known to be HIV positive, the risk of transmission is increased in the late stage of infection (16). The patient should be asked about evidence of a high viral titer in his or her sexual contact: signs of AIDS, low CD4 count, or primary infection. Conversely, the risk of transmission is decreased by about 50% if the contact is on antiretroviral therapy (16).

The type of sex also plays a major role in the likelihood of transmission. Receptive intercourse is associated with much higher rates of transmission. For women having intercourse with HIV-infected men, the probability of acquiring HIV has been estimated to be about 0.001 (17). For men having receptive intercourse with HIV infected men, the probability is about 0.02 (17). Insertive intercourse has been associated with a much lower risk. The probability of acquiring HIV from either heterosexual or homosexual insertive intercourse is estimated at 0.0006 (17). Of course, condoms are highly effective in decreasing HIV transmission.

How Effective is Post-Exposure Prophylaxis?

There is little data on the efficacy of prophylaxis after sexual exposure. Most experts have extrapolated data derived from health care workers exposed to contaminated blood through needle sticks. In that setting, zidovudine has been reported to decrease the risk of transmission by about 79% (18,19). The efficacy of regimens containing more than one drug, which are currently recommended when administering prophylaxis, is unknown.

What Are the Risks Involved in Taking Post-Exposure Prophylaxis?

Common side effects of zidovudine and lamivudine include malaise, headache, and nausea. If patients discontinue the regimen early because of unpleasant side effects, they may increase the risk of developing viral resistance (20). Another major concern regarding the use of PEP is that patients may be more likely to engage in high-risk behavior if they assume incorrectly that PEP will prevent infection.

Should the Patient Receive Post-Exposure Prophylaxis?

All patients at risk should be informed about the risks and benefits of PEP, although it is not indicated in all situations (21). PEP (with zidovudine and lamivudine) has been recommended in the case of a high-risk exposure (unprotected anal, vaginal, or receptive oral intercourse with a partner who is or is likely to be HIV infected) that occurs as a singular event or involves someone who intends to stop further high-risk behavior (20,22). Some patients may not meet these criteria yet still wish to receive PEP.

It is also important for patients to recognize that cost may represent a significant barrier. Some insurance companies may not pay for PEP, and the estimated cost of a 28-day course of triple therapy may be in excess of $1,000, and lab and follow-up tests may add an additional $500.

WHAT TO TELL THE PATIENT

It is not appropriate to simply "call in a prescription" for antiretroviral therapy over the telephone. This is an extremely important and complex decision. If the patient decides to take PEP, he or she should seek medical attention in an emergency department or physician's office as soon as possible.

Although data are scant, it appears prudent to administer PEP within hours of the exposure, if possible (21). Patients should have their blood drawn for HIV testing before initiating therapy, and informed consent should be obtained. Patients should receive counseling about the HIV test and the risks and benefits of prophylaxis. Patients should be tested and screened for other sexually transmitted diseases and referred for follow-up testing and continued preventive services (23). If the partner's HIV status is unknown, he or she should also be tested to confirm HIV status and viral load, if necessary. If the partner's antibody test is negative and

PATIENT WHO CALLS WITH SEXUAL EXPOSURE TO HIV+

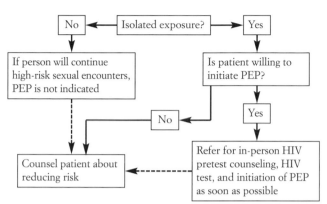

there is no detectable virus, then the exposed person can be reassured that transmission was highly unlikely.

WHAT TO DOCUMENT

- Patient's description of the high-risk exposure and whether it took place within the last 72 hours
- HIV risk-factor assessment
- Whether the exposure was an isolated event
- If indicated, whether patient is willing to start PEP: if yes, when the patient will be seen for evaluation
- Whether risk counseling was done
- Follow-up plan

WHAT IS A GOOD SOURCE FOR UP-TO-DATE POST-EXPOSURE PROPHYLAXIS RECOMMENDATIONS?

The National Clinicians' Post-Exposure Prophylaxis Hotline (PEPLine) offers free consultation for clinicians treating patients exposed to HIV. The telephone number is 888-HIV-4911, and the staff is available 24 hours a day, 7 days a week.

HEPATITIS B PROPHYLAXIS

Epidemiology

Approximately 200,000 to 300,000 people become infected with hepatitis B each year in the United States, and there are more than 1 million chronic carriers (24). The prevalence of hepatitis B carriers has been reported to be as high as 20% in specific high-risk populations such as chronic renal dialysis patients and intravenous drug users (25).

With improved testing of blood donors, the major mechanism of hepatitis B transmission in the United States is no longer via blood transfusions. The most common means of transmitting hepatitis B is through sexual contact; hepatitis B vaccination is now recommended for all young adults and particularly for individuals with multiple sexual contacts (24). The purpose of evaluating a patient for possible HBV exposure over the phone is to prevent infection and its long-term complications such as hepatoma and cirrhosis of the liver.

How Does One Evaluate the Risk of Acquiring Hepatitis B Through Sexual Contact?

The likelihood of transmitting HBV depends on the hepatitis B status of the contact, the nature of the sexual exposure, and the immunization status of the patient. If the partner has active hepatitis B infection or is a carrier with hepatitis B surface antibody (HBsAg[+]), there is an increased risk of transmission. If the partner's status is unknown but he or she is available and willing to be tested, prophylaxis in the patient can be delayed until the results are available. However, if results will not be available for 14 days or longer, it is probably not advisable to wait for testing to initiate prophylaxis.

The immunization status of the patient should also be assessed. A documented response of HBsAb positive generally confers protection for more than 7 years. If there is uncertainty about the response or if the exposure is high risk, the patient should have titers checked. Further intervention is probably not necessary if HBsAb levels are adequate.

PATIENT WHO CALLS WITH POSSIBLE EXPOSURE TO HEPATITIS B VIRUS
WITHIN PAST 14 DAYS

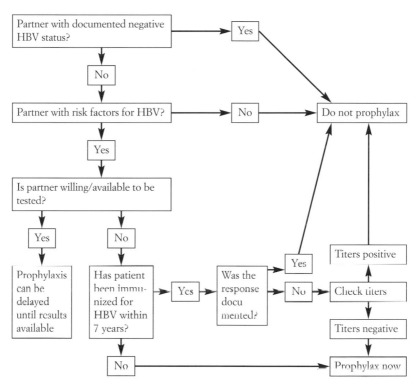

How Effective is Post-Exposure Prophylaxis?

Both active and passive immunization are recommended in the acute setting. Passive immunization is provided by hepatitis B immune globulin (HBIG), which is prepared from plasma containing high amounts of hepatitis B surface antibody. HBIG may decrease the risk of transmission by up to 75% if taken within 2 weeks of exposure (26,27).

The HBV vaccine has also been demonstrated to decrease the risk of transmission acutely, although published data are available only for perinatal exposure (27). Because people who are concerned about recent sexual exposure to HBV are at increased risk of exposure in the future, the HBV vaccine should be routinely administered to all such patients. Hence, PEP for HBV exposure should consist of both HBIG and the initiation of an HBV vaccine series.

WHAT TO TELL THE PATIENT

The patient should be advised to make an appointment with his or her physician and to bring the partner, if possible, for testing as well. Explain to the patient that the PEP regimen for hepatitis B is effective for as long as 14 days after exposure, so unless the exposure was almost 14 days ago, it is not necessary to go the emergency room.

WHAT TO DOCUMENT

- Patient's description of the high-risk exposure and whether it took place within the last 14 days
- HBV risk-factor assessment
- Partner's HBV status, if known
- Whether patient received HBV vaccine
- Whether risk counseling was done
- Follow-up plan

REFERENCES

1. **Trussell J, Steward F.** The effectiveness of postcoital contraception. Fam Plan Perspect. 1992;24:262-4.
2. **Glasier A.** Emergency postcoital contraception. N Engl J Med. 1997;337:1058-64.
3. **Wilcox A, Weinberg C, Baird D.** Timing of sexual intercourse in relation to ovulation. N Engl J Med. 1995;333:1517-21.
4. **Barrett J, Marchall J.** The risk of contraception on different days of the menstrual cycle. Popul Stud. 1969;23:455-61.
5. **Grou F, Rodriguez I.** The morning after pill—how long after? Am J Obstet Gynecol. 1994;171:1529-34.
6. **Trussell J, Rodriguez G, Ellertson C.** New estimates of the effectiveness of the Yuzpe regimen of emergency contraception. Contraception. 1998;57:363-9.
7. **Simpson J, Phillips O.** Spermicides, hormonal contraception, and congenital malformation. Adv Contracept. 1990;6:141-67.
8. **Bracken M.** Oral contraception and congenital malformation in off-spring: a review and meta-analysis of prospective studies. Obstet Gynecol. 1990;76:552-7.
9. ACOG Practice Patterns. Emergency Contraception. Washington, DC: ACOG, 1996.
10. **Stewart F.** Promoting emergency contraception. Hosp Prac. 1998;33:61-3, 67-9, 73-5.
11. **Task Force on Postovulatory Methods of Fertility Regulation.** Randomised controlled trial of levonorgestrel versus the Yuzpe regimen of combined oral contraceptives for emergency contraception. Lancet. 1998;352:428-33.

12. **Women's Capital Corporation.** A new generation of emergency contraception has arrived: FDA approves progestin-only emergency contraceptive, 1999.
13. **Abramowicz M.** An emergency contraceptive kit. Med Lett. 1998;40:102-3.
14. **Campbell K.** Methods of monitoring ovarian function and predicting ovulation: summary. Res Front Fertil Regul. 1985;3:1-16.
15. **Prevention CfDCa.** HIV/AIDS Surveillance Report. Atlanta; 1999:3-33.
16. **Royce R, Sena A, Cats W, Cohen M.** Sexual transmission of HIV. N Engl J Med. 1997;336:1072-8.
17. **Pinkerton S, Holtgrave D, Bloom F.** Cost-effectiveness of post-exposure prophylaxis following sexual exposure to HIV. AIDS. 1998;12:1067-78.
18. Case-control study of HIV seroconversion in health-care workers after percutaneous exposure to HIV-infected blood. MMWR. 1995;44:929-33.
19. **Cardo D, Culver D, Ciesielski C.** A case-control study of HIV seroconversion in health care workers after percutaneous exposure. N Engl J Med. 1997;337:1485-90.
20. **Katz M, Gerberding J.** Postexposure treatment of people exposed to human immunodeficiency virus through sexual contact or injection-drug use. N Engl J Med. 1997;336:1097-1100.
21. **Lurie P, Miller S, Hecht F, et al.** Postexposure prophylaxis after nonoccupational HIV exposure. JAMA. 1998;280:1769-73.
22. **Katz M, Gerberding J.** The care of persons with recent sexual exposure to HIV. Ann Intern Med. 1998;128:306-12.
23. **American Medical Association.** A Physician's Guide to HIV Prevention. Chicago: AMA; 1996.
24. **United States Preventive Services Task Force.** Guide to Clinical Preventive Services. Baltimore, 1996:/3-119.
25. **Dienstag J, Isselbacher K.** Acute viral hepatitis. In: Fauci A, Braunwald E, Isselbacher K, et al, eds. Harrison's Principle's of Internal Medicine. New York: McGraw-Hill, 1998:1677-92.
26. **Redeker A, Mosley J, Gocke D, et al.** Hepatitis B immune globulin as a prophylactic measure for spouses exposed to acute type B hepatitis. N Engl J Med. 1975;293:1055-9.
27. **Immune Practices Advisory Committee.** Protection against viral hepatitis. MMWR. 1990;39:1-22.

14

···············

DOMESTIC VIOLENCE

················

Jeanne McCauley, MD, MPH

KEY POINTS

- Domestic violence is common and is associated with injury and many physical and psychological problems.
- Patients will rarely call with a chief complaint of domestic violence, so clinicians must be alert to the possibility of abuse.
- Clinicians should try to ensure that patients are in a private setting before asking about the possibility of abuse.
- It is critical to assess a patient's current level of safety in addition to his or her medical concerns.
- A clinician does not have to be an expert in domestic violence to be effective. An empathetic attitude and knowledge of local domestic violence resources can be extremely helpful to the patient.
- Many survivors report that validation of their suffering, even from one person, helped restore self-esteem.
- 1-800-799-SAFE is a national 24-hour, 7-day-a-week, toll-free hotline available for domestic violence victims.

Domestic violence affects 2 to 4 million women in the United States per year; one third of women presenting to primary care offices will have suffered abuse in their lifetime. In addition to injury, domestic violence is associated with many physical and psychological problems. Although patients rarely call with a primary complaint of domestic violence, clinicians must be alert to the possibility of abuse. Clinicians should try to ensure that patients are in a private setting before they ask about the possibility of abuse. It is critical to assess a patient's current level of safety (e.g., "Do you think your life is in danger right now?") in addition to her medical concerns. A clinician does not have to be an expert in domestic violence to be effective. An empathetic attitude and knowledge of local domestic violence resources can be extremely helpful to the patient. This chapter will focus on the following questions:

1. Is this patient being abused?
2. Is this patient's life in danger?
3. How can the physician screen and treat domestic violence over the telephone?

BACKGROUND

Epidemiology

Domestic violence is the threat or use of physical force against a person by a relative or intimate who may or may not be living in the same household with the abused person. It is estimated that 2 to 4 million women in the United States are physically abused each year; domestic violence may occur in as many as one in every four U.S. families (1). The prevalence of current domestic violence in adult women presenting to primary care settings has been estimated to range from 5% to 23%; the lifetime adult prevalence ranges from 8% to 44% (1-6). Although domestic violence is a significant risk for women of any age race, age, or socioeconomic background, domestic violence tends to be more common in young patients (the highest prevalence being in women less than age 25) and in women who are single or separated, and some experts feel that it is most common in women with low incomes (2,4,5).

Men can also be victims of domestic violence, but in most cases the prevalence and severity is felt to be lower than for women. There is little medical research on health problems of men experiencing violence.

Utility of Early Diagnosis

Telephone identification of women who are victims of domestic violence can be the first step toward obtaining help and improving patient safety and health. Risks of diagnosing domestic violence over the telephone exist, however, in terms of the caller's safety. This chapter will provide information on identifying when domestic violence should be addressed over the telephone and when it should not.

WHAT KIND OF HEALTH PROBLEMS DO DOMESTIC VIOLENCE VICTIMS HAVE?

Abuse, whether experienced currently, in past adulthood, or in childhood or adolescence, is associated with a variety of physical and psychological symptoms and probably increased health care utilization (Tables 14-1 and 14-2) (2,7-15). The presenting symptoms of patients who call and have experienced domestic violence have not been specifically studied; however, it is useful to review common ambulatory presentations.

Table 14-1   Physical Symptoms in Women Experiencing Current Domestic Violence

• Loss of appetite	• Abdominal or stomach pain
• Bruises (frequent or serious)	• Breast pain
• Nightmares	• Headaches (frequent or serious)
• Vaginal discharge	• Urinary symptoms
• Eating binges or self-induced vomiting	• Chest pain
• Diarrhea	• Sleeping problems
• Broken bones, sprains or cuts	• Shortness of breath
• Pelvic or genital pain	• Constipation
• Fainting or passing out	

Table 14-2 Psychological Symptoms in Women Experiencing Current Domestic Violence

• Anxiety	• Attempted suicide
• Depression	• Considered suicide in past 7 days
• Somatization	• Post-traumatic stress disorder
• Increased interpersonal sensitivity	• Drug or alcohol abuse

Physical Problems

Abused women suffer from higher levels of gastrointestinal disorders, headaches, chronic pain, pelvic pain, and multiple somatic complaints (Table 14-1) (2,7-15). Prevalence of abuse during pregnancy may be as high as 17% (16). In one study of physical symptoms of female current domestic violence victims in a primary care setting, abused women had more physical symptoms in the previous 6 months than women who were not experiencing current abuse (mean 7.3 vs. 4.3). The specific physical symptoms more common in the currently abused group included headaches, chest pain, shortness of breath, sprains, fractures, bruises, abdominal and pelvic pain, urinary frequency, and vaginal discharge (2). These studies suggest that violence is associated with many different physical symptoms. It is unclear whether some of the physical problems are caused by the trauma itself, by a heightened sensitivity to pain because of early neural damage, or by psychological distress associated with the trauma (17).

Psychological Symptoms

In primary care and subspecialty settings, a history of violence either currently or in the past has been found to be significantly associated with depression, anxiety, low self-esteem, somatization, post-traumatic stress disorder, current and past drug and alcohol abuse, and suicide attempts (Table 14-2) (2,15,18-22).

Medical Utilization

Although there have been few studies of health care utilization in battered women, the present literature to date suggests that abused women seek health care with disproportionate frequency (22-25). In one study of the health effects of criminal victimization, the majority of the assaults and rapes were committed by an intimate or acquaintance of the victim. Although the study was limited by a low response rate, outpatient physician visits (excluding psychotherapy) increased 15% to 24% in the year after the crime. In this study, victimization was the most powerful predictor of physician visits and outpatient costs (24). In a study of battered women in a primary care setting, currently abused women were more likely to have visited an emergency room in the 6 months before presentation when compared to not currently abused controls (2). In a 1996 report from the National Institute of Justice in which the rape count rate may be "tenuous and low," the annual medical costs for adult domestic violence were estimated to be $1.8 billion; the total annual societal costs (including tangible property loss and impact on the quality of life) was estimated to be $67 billion (25).

IF DOMESTIC VIOLENCE IS SO COMMON, WHY IS IT RARELY A CHIEF COMPLAINT? WHY WOULD A WOMAN NOT JUST LEAVE?

Fewer than one in three abused patients have ever discussed abuse with a physician or other health professional (Table 14-3) (2,26,27). Most patients do not spontaneously volunteer a history of abuse because of shame, denial that they are "abused" even when the physical harm is severe, fear of retaliation from the abuser, financial dependence on the abuser, fear of legal custody issues regarding their children, fear of repercussions from other family members, emotional commitment to the abuser, helplessness, and hopelessness. Some women are physically prevented by their abusers from seeking routine or emergency health care (26,27).

WHY DO MOST PHYSICIANS NOT ROUTINELY ASK ABOUT DOMESTIC VIOLENCE?

The majority of physicians do not screen their patients for abuse because of time pressures and lack of knowledge about prevalence, presentation,

---

Table 14-3 Barriers to Physician and Patient Discussion of Abuse

---

Physician Barriers

Lack of time

Fear of offending the patient by asking about abuse

"Too close for comfort"

Powerless to help or "fix" the problem

Loss of physician control over patient's decision to leave or seek help

Lack of continuity with patient

Lack of education about detection and treatment

Embarrassment in screening patients of high socioeconomic status

Patient Barriers

Shame, embarrassment, fear of reaction of others (including physicians)

Denial

Fear of repercussions from abuser

Lack of financial resources to get food, housing, or medical care

Fear of police involvement

Fear that family will be separated

---

and treatment (Table 14-3) (28,29). Too often, physicians underestimate the prevalence of domestic violence and doubt that a patient could be abused if she is of the middle or upper class or if she is well educated. Some physicians blame the patient, feeling that she precipitated the abuse or is weak if she fails to "just leave."

However, experts would respond that many abused women do present to office practices or call for medical advice with multiple physical problems and psychological distress and that clinicians are in a unique position to deal with the physical and psychological problems associated with abuse.

How Will an Abused Woman Present When She Calls?

Women experiencing domestic violence may call primary care clinicians with a variety of complaints, including:

- Injury (some patients may delay seeking help for several days because their abusers have physically prevented them from calling sooner; also, the history the patient gives may not fit the pattern of her injuries.)
- Depression/suicidal ideation
- Panic attack symptoms
- New physical complaint
- Exacerbation of an existing medical condition
- Domestic violence itself

To date, there have been no studies on which complaints are most common and how frequent these complaints are. If the women who call for help are similar to the women who present in person, complaints of domestic violence itself will be rare, and accidents, depression and physical complaints will be much more common.

WHAT ARE THE BENEFITS AND RISKS OF SCREENING A WOMAN ON THE TELEPHONE FOR DOMESTIC VIOLENCE?

Benefits of screening a woman on the telephone for domestic violence include detection of abuse, which can be the first step toward obtaining help and improving patient safety and health.

Risks include increased harm to patients through several mechanisms. If an abuser is listening or uses caller ID to determine whom the patient has been calling, this can put the patient at increased risk. In addition, if a patient is sent to an emergency room or other clinician and then encounters a cold or judgmental attitude, she may suffer from secondary stigmatization (27,30). Because many abused patients feel intense shame and stigma about the violence, there may be denial or concealment of the abuse. Secondary stigmatization is additional shame brought on by being ignored, blamed, rejected, or "judged" by a clinician (30). This secondary stigmatization can add a severe amount of additional psychological stress to the patient. It is important for the physician to call the emergency room to inform the emergency room physicians and nurses that the patient is a domestic violence victim.

The physician can decrease risk by:

- Asking if the patient is in a private area
- Supporting and reassuring the patient if she reveals a history of domestic violence
- Arranging a follow-up appointment with the patient either by telephone or in person in a safe environment

Generally, it is better to perform routine screening for domestic violence in person in a confidential office setting. A patient should be screened if a clinician feels that a "yes" answer to the question "Have you been physically hurt or has anyone forced you to have any sexual activity against your will?" will effect immediate disposition.

WHICH PATIENTS SHOULD BE SCREENED OVER THE TELEPHONE?

Telephone screening is indicated when there is risk of severe and immediate harm. For example:

- A patient who describes even "mild" trauma or bruises, a substance-abusing partner, or multiple physical or psychological problems
- A patient who calls with a gynecological complaint that may be related to a rape. In many cities, only certain emergency departments perform "rape exams" because they have special forensic equipment. If a rape has occurred within 72 hours, the patient should be sent to one of those centers. The rape victim should be urged to seek emergency department evaluation as quickly as possible and to not shower or change because this can destroy forensic evidence. Proper collection and labeling of bodily fluids is critical in rape prosecution.
- A depressed patient
- Any patient with a known history of abuse with new or exacerbated medical complaints
- Any patient who sounds afraid

HOW SHOULD THE PHYSICIAN SCREEN A PATIENT FOR DOMESTIC VIOLENCE OVER THE TELEPHONE? HOW CAN THE PHYSICIAN DETERMINE IF A WOMAN IS IN CURRENT DANGER, AND WHAT ADVICE SHOULD BE GIVEN?

A clinician is a part of a team and need not "know all and do all" to be effective in screening and treating abuse. The therapeutic importance of a clinician affirming that no one "deserves" to be abused, or that a patient is not "stupid" or "weak" for being in an abusive relationship, cannot be underestimated. Many survivors report that this validation, even from one person, helped them restore their self-esteem. If a clinician can do no more than reassure the patient and provide appropriate referral and follow-up, a real service has been performed.

Before screening, be sure to ask if the patient is in a safe place to talk: "Are you in a private enough area to talk?" If no, "Can I call you or can you call me in a more private area?"

The acronym ASSERT (Ask, Sympathize, Safety, Educate, Record, Triage/Treat) serves as a practical clinical tool for the clinician who is assessing a patient for domestic violence (Table 14-4).

Ask directly: Most patients do not call with a chief complaint of "domestic violence or abuse," and direct questioning should be used once the decision to screen has been made. It is important to not use the term "abuse" in the screening question because many abused patients are in denial, even when the physical injury is severe. For example: "Violence is so common that I frequently ask about it. Have you been physically hurt or has anyone forced you to have any sexual activity against your will?"

Sympathize: Because many abused patients are afraid, alone, and feel "stupid" for tolerating the abuse, it is important that clinicians offer an immediate empathetic statement of support. A sympathetic tone of voice is critical; if a patient perceives an indifferent, distant, or uncaring attitude from the clinician, she may fear to disclose abuse (27,30). For example: "This must be very hard for you to discuss. No one deserves to be hurt. I want you to know that I am concerned about you and will do what I can to help you."

**Table 14-4   Screening and Treatment of Domestic Violence
Over the Telephone**

Before screening, be sure to ask if the patient is in a safe place to talk. The acronym **"ASSERT"** (**Ask, Sympathize, Safety, Educate, Record, Triage/Treat**) serves as a practical tool for the clinical assessment of a patient for domestic violence.

- **A—Ask directly:** "Violence is so common that I frequently ask about it. Have you been physically hurt or has anyone forced you to have any sexual activity against your will?

- **S—Sympathize:** "This must be very hard for you to discuss. No one deserves to be hurt. I want you to know that I am concerned about you and will do what I can to help you."

- **S—Safety:** "If you stay where you are, do you feel that you will be safe today?"

- **E—Educate:** Tell the caller about the closest shelters and the services that they provide. As of 1999, a national 24-hour, 7-day-a-week, toll-free hotline became available. The number is 1-800-799-SAFE. Shelters are valuable resources for information on protection orders.

- **R—Record:** Urge the caller to let you document the abuse in the telephone triage note and add it to her medical chart. Such documentation may help her get protection orders or child custody.

- **T—Triage/Treat:** Ask the caller about any injuries and any medical, psychological, or substance abuse problems associated with the abuse that would require immediate evaluation.

Some patients may "test" the clinician to assess concern (invitational disclosure); in this situation, patients may hint that something is wrong in their personal life or say they are under a lot of "stress" but will not disclose unless directly "invited to" by a direct question from the clinician (27,30). Avoid implying that the patient did anything to provoke the abuse; avoid minimizing the abuse; and avoid telling the patient she should "just leave." Domestic violence is a complex problem, and frequently a patient lacks the energy to leave her abuser because of depression, low self-esteem, absence of financial or family support, and fears of losing custody of children. Additionally, some experts feel that an abused person is at greatest risk for homicide in the immediate period after leaving an abuser.

Safety: Patients should be asked if they feel that they are in current danger of trauma or death. For example: "If you stay where you are, do you feel that you will be safe today? "If the patient feels that she is in current danger, urge her to call a friend, relative, or shelter for a place to stay. Many patients will state that it is safe for them to stay where they are.

Past history of severe abuse, firearms in the house, and a substance-abusing or intoxicated partner are risk factors for severe injury or death in patients. Suspicion of risk for high-level violence should increase efforts to persuade the patient to leave the area and get a protection order as soon as possible. In addition, a depressed woman should be screened for suicidal ideation. *If a patient's current condition is serious or if the abuser is an immediate threat to the patient, the physician should call 911.*

Educate: Tell the caller about the closest shelters and the services that they provide. In addition to shelter care, some shelters offer therapy groups, legal advice, or therapy groups for abusers. Social workers and mental health professionals may also be knowledgeable about treatment options and local resources. As of 1999, a national 24-hour, 7-day-a-week, toll-free hotline became available: 1-800-799-SAFE. Shelters are valuable resources for information on protection orders.

Record: Patients should be urged to have their abuse documented in the telephone triage note and added to their medical chart. Such documentation may help them get protection orders or child custody. The complaints should be as close as possible to the patient's actual words. Never state "The patient alleges that she/he experienced...." The word "alleges" is pejorative and implies that the clinician doubts the patient's story. Patients should be assured that their information will not be discussed with other callers. It is important to schedule an appointment for the patient to be seen as quickly as possible in order to document a physical exam and any sign of trauma.

Reporting to police/protective services varies by state. Currently most states do not require mandatory reporting of domestic violence.

Triage/**T**reat: Treat any injuries and medical, psychological, or substance-abuse problems associated with the abuse that would require immediate evaluation. Any patient who fulfills medical or psychological criteria for immediate emergency room or office evaluation should be advised to go to the emergency room. A follow-up appointment or telephone call should be offered and provided. Try to arrange a safe place and time to call the patient because some patients are prevented by their abusive partners from returning calls or keeping appointments.

PATIENT WHO CALLS WITH SYMPTOMS SUGGESTING DOMESTIC VIOLENCE

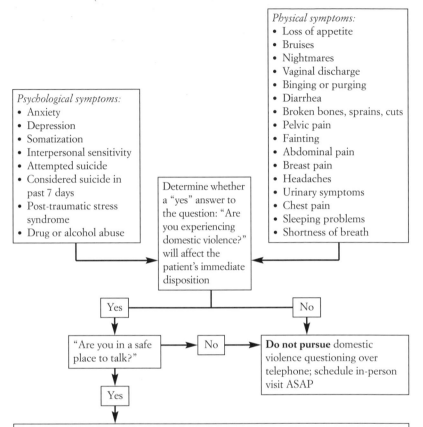

Psychological symptoms:
- Anxiety
- Depression
- Somatization
- Interpersonal sensitivity
- Attempted suicide
- Considered suicide in past 7 days
- Post-traumatic stress syndrome
- Drug or alcohol abuse

Physical symptoms:
- Loss of appetite
- Bruises
- Nightmares
- Vaginal discharge
- Binging or purging
- Diarrhea
- Broken bones, sprains, cuts
- Pelvic pain
- Fainting
- Abdominal pain
- Breast pain
- Headaches
- Urinary symptoms
- Chest pain
- Sleeping problems
- Shortness of breath

Determine whether a "yes" answer to the question: "Are you experiencing domestic violence?" will affect the patient's immediate disposition

Yes

No

"Are you in a safe place to talk?"

No

Do not pursue domestic violence questioning over telephone; schedule in-person visit ASAP

Yes

Follow the ASSERT model:
1. **A**sk: Ask about domestic violence directly.
2. **S**ympathize: Make an empathetic statement of support.
3. **S**afety: Ask whether the patient feels safe in her current environment.
4. **E**ducate: Provide information on shelters and hotlines.
5. **R**ecord: Urge patient to allow the information to be recorded in the medical chart.
6. **T**riage/Treat: Arrange proper follow-up, depending on acuity of situation.

WHAT TO DOCUMENT

- Physical and/or psychological symptoms (as in Tables 14-1 and 14-2) described by the patient, as close as possible to her own words
- Patient's sense of safety

- Information on hotlines and/or shelters that was given
- Granting of patient permission to document the encounter
- Physician instructions to the patient and whether patient agreed to follow instructions
- Date of scheduled in-person visit

REFERENCES

1. **Novello AC.** From the Surgeon General: A medical response to domestic violence. JAMA. 1992;267:31-2.
2. **McCauley JM, Kern DE, Kolodner K, et al.** The battering syndrome: prevalence and clinical characteristics of domestic violence in primary medical care internal medicine practices. Ann Intern Med. 1995;123:737-46.
3. **Rath GD, Jarratt LG, Leonardson G.** Rates of domestic violence against women by male partners. J Am Board Fam Pract. 1989;2:227-33.
4. **Ellicot BA, Johnson MM.** Domestic violence in a primary care setting. Arch Fam Med. 1995;4:113-19.
5. **Hamberger LK, Saunders DG, Hovey M.** Prevalence of domestic violence in community practice and rate of physician inquiry. Fam Med. 1992;24:283-7.
6. **Bullock L, McFarlane J, Bateman LH, Miller V.** The prevalence and characteristics of battered women in a primary care setting. Nurse Pract. 1989;14:47-55.
7. **Drossman DA, Lesserman J, Nachman G, et al.** Sexual and physical abuse in women with functional or organic gastrointestinal disorder. Ann Intern Med. 1990;113:828-33.
8. **Harber JD, Roos C.** Effects of spouse abuse in the development and maintenance of chronic pain in women. Adv Pain Res. 1985;9:889-95.
9. **Reiter RC, Shakerin LR, Gambone DO, Milborn AK.** Correlation between sexual abuse and somatization in women with somatic and nonsomatic chronic pelvic pain. Am J Obstet Gynecol. 1991;165:104-9.
10. **Walter E, Katon W, Harrop-Griffiths J, et al.** Relationship of chronic pelvic pain to psychiatric diagnoses and childhood sexual abuse. Am J Psych. 1988;145:75-80.
11. **Walker E, Katon WJ, Hanson J, et al.** Medical and psychiatric symptoms in women with childhood sexual abuse. Psychosom Med. 1992;54:658-64.
12. **Briere J, Zaidi LY.** Symptomatology associated with childhood sexual victimization in a nonclinical adult sample. Child Abuse Neglect. 1988;12:51-9.
13. **Mullen PE, Romans-Clarkson SE, Walton CA, Herbison GP.** Impact of sexual and physical abuse on women's mental health. Lancet. 1988;12:51-9.
14. **Schei B, Bakketeig LS.** Gynecological impact and sexual and physical abuse by spouse: a study of intrafamily conflict and violence: the conflicts tactics (CT) scale. J Marriage Fam. 1979;41:75-88.
15. **McCauley J, Kern DE, Kolodner K, et al.** Clinical characteristics of women with a history of child abuse: unhealed wounds. JAMA. 1997;227:1362.
16. **Newberger EH, Barkan JE, Lieberman ES, et al.** Abuse of pregnant women and adverse birth outcomes. JAMA. 1992;267:2370-2.

17. **Scarinci K, McDonald-Haile J, Bradley LA, et al.** Altered pain perception and psychosocial feelings among women with GI disorders and a history of abuse. Am J Med. 1994;97:108-18.
18. **Briere J, Zaidi LY.** Sexual abuse histories and sequelae in female psychiatric emergency room patients. Am J Psych. 1989;146:1602-6.
19. **Jaffe P, Wolfe DA, Wilson S, Zak L.** Emotional and physical health problems of battered women. Can J Psychiatry. 1986;626-9.
20. **Brown GR, Anderson B.** Psychiatric morbidity in adult inpatients with childhood history of sexual or physical abuse. Am J Psychiatry. 1991;148:55-61.
21. **Moeller TP, Bachmann GA, Moeller J.** The combined effects of physical, sexual, and emotional abuse during childhood: long-term health consequences for women. Child Abuse Neglect. 1993;17:623-40.
22. **Plichta S.** The effects of women abuse on health care utilization and health care status: a literature review. WHI. 1992;2:154-63.
23. **Bergman B, Brismar B.** A 5-year follow-up study of 118 battered women. Am J Public Health. 1991;81:1486-9.
24. **Koss MP, Koss PG, Woodruff WJ.** Deleterious effects of criminal victimization on women's health and medical utilization. Arch Intern Med. 1991;151:342-7.
25. **Miller TR, Cohen MA, Wiersema B.** Victim costs and consequences: a new look. Report to the National Institute of Justice 1996 (NCJ 155282).
26. **Loring MT, Smith RW.** Health care barrier and interventions for battered women. Pub Health Rep. 1991:109:328-38.
27. **McCauley J, Yurk RA, Jenckes MW, et al.** Inside Pandora's box: abused women's experiences with clinicians and health services. J Gen Int Med. 1998;13:549-55.
28. **Ferris LE.** Canadian family physicians' and general practitioners' perceptions of their effectiveness in identifying and treating wife abuse. Med Care. 1994;32:1163-72.
29. **Sugg NK, Inui T.** Primary care physicians' response to domestic violence-opening Pandora's box. JAMA. 1992;267:3157-60.
30. **Limandri BJ.** Disclosure of stigmatizing conditions: the disclosure's perspective. Arch Psychiatric Nursing. 1989;2:69-78.

15

DEPRESSION AND SUICIDALITY

William H. Salazar, MD

KEY POINTS

- Many patients with underlying depression will present initially with somatic complaints; patients calling with multiple or vague somatic complaints are at high risk for depression.
- Most patients who attempt suicide will seek out medical attention before their suicide attempt, but their complaints may be physical in nature.
- Given the high risk of suicide in depressed patients, physicians should have a low threshold for ruling out suicidal ideation over the phone.
- Suicidality should be ruled out with direct questioning.
- Patients with possible active suicidality should be referred to an emergency department.
- Some states require that physicians contact the police in certain settings involving possible suicidality.
- Patients at the greatest risk for death from suicide are those with a history of suicide attempt in the past and those with a specific plan, especially if it is a violent method (e.g., gunshot).
- In the possibly suicidal patient, always obtain the full name, telephone number, and address early in the telephone call.

Continued

- If an actively suicidal patient refuses to go to an emergency facility, the physician should contact the police and a psychiatrist at the emergency facility.
- Depressed patients who are not suicidal should be encouraged to see their primary care physician soon to begin treatment.
- Patients with a past history of major depression who were responsive to antidepressants and are now experiencing a recurrence of depression but are no longer taking antidepressants may benefit from immediately restarting their previous medication.

Depression is a very common and disabling illness. On the telephone, physicians may encounter calls from patients with either new-onset depression or previously diagnosed depression. As is the case with office visits to primary care physicians, many cases of long-standing depression will likely be undiagnosed.

Depressed patients may initially conceal their depression from a physician, calling instead about somatic complaints such as headaches, fatigue, and abdominal pain. Because of the high risk of suicide in depressed patients (15%) (1), physicians should err on the side of caution with emergency department referral. Only patients who are actively suicidal should be referred to the emergency department. Although the evidence is inconclusive, some suicides among depressed patients may be preventable with psychiatric treatment (2).

The primary reasons for considering the diagnosis of depression over the telephone are to manage possible suicidality and to help the patient get appropriate treatment. This chapter will focus on determining if the patient has major depression and whether the patient should be hospitalized for suicidality.

BACKGROUND

Epidemiology

One in six people will suffer from major depression at some point during their lives (lifetime prevalence of 17%). In primary care settings, the prevalence of depression may be as high as 12% (3), similar to the prevalence of diabetes mellitus and hypertension. Like these medical conditions, depression is a chronic medical disorder (the likelihood of recurrence after the first episode is 50% [4]) that can produce profound functional limitation and morbidity.

Unfortunately depression continues to be underdiagnosed and undertreated by primary care physicians (5), with about half of the psychiatric comorbidity in patients visiting their primary care physician going unrecognized (6). Depressed patients' chief complaints are rarely of a psychological nature, with the majority describing somatic symptoms (41%), pains (37%), and fatigue and sleep problems (12%) (7). In addition, the high comorbidity rates with somatoform and anxiety disorders, also frequently dominated by somatic symptoms, may contribute considerably to the poor recognition and diagnosis of depressive disorders in primary care settings.

One of the most important and ongoing responsibilities primary care physicians have with all depressed patients is assessing their risk of suicide (7). Most suicidal patients have a psychiatric condition, most commonly major depression (2). In untreated depression, 80% of patients express suicidal ideation, 20% to 40% exhibit suicidal behavior, and up to 15% succeed in killing themselves (8).

Telephone calls in which the stated reason for calling is psychiatric seem to be infrequent, although there may be a detection bias. One study found that only 0.6% of the calls were for psychiatric-related problems, compared to 15.4% for infectious diseases and 9.4% for gastrointestinal problems (9). It is quite possible, however, that many patients with underlying psychiatric concerns did not express them to the doctor for the reasons given above.

Utility of Early Diagnosis

Major depression responds well to treatment with medication and psychotherapy, and there is evidence that early diagnosis and treatment improves response rates (2). Active suicidality is usually managed with hospitalization and psychiatric treatment, but the evidence that this intervention reduces the patient's risk of suicide remains inconclusive. The reasons for this are complex, but may be partly because of the relative infrequency of completed suicide compared to the larger number of patients with suicidal ideation. This makes demonstration of efficacy of preventive measures difficult (2). Nevertheless, there is no evidence that psychiatric treatment is clearly not effective, and inpatient treatment remains the standard of care. It seems reasonable to presume that early detection and intervention is superior to no treatment and superior to waiting for suicidality to become manifest in other ways, such as an unsuccessful suicide attempt.

Suicidal patients frequently contact a physician before committing suicide. Although there are no data on how many of these patients call their physician, 50% of patients who commit suicide visit a primary care physician in the month before their death (2). Clearly, suicidal patients are eager to make contact with their physicians. Some of these patients may be more comfortable making this contact on the telephone.

WHEN SHOULD A PATIENT ON THE TELEPHONE BE SCREENED FOR DEPRESSION?

Screening for depression can be relatively quick and straightforward. Given the significant morbidity and mortality associated with depression, the physician should have a low threshold for screening for this condition.

There is no rule about which callers are more or less likely to have undiagnosed depression or suicidality. As described in the Background section, depressed patients usually give a physical complaint as their primary reason for speaking with a primary care physician (10), so a physician cannot limit his or her suspicion for depression to patients whose chief complaint is emotional in nature. A variety of clinical settings suggest depression (Table 15-1). Physicians should be concerned about depression when the clinical presentation does not make physiological sense, when the patient *seems* sad, or when the interview results in the *physician* feeling sad.

Patients vary in their openness to disclosing emotional symptoms over the telephone. One small study (11) showed increased openness to discussing psychiatric symptoms over the phone compared to the office

---

Table 15-1   Settings in Which One Should Assess for Depression

- Chief complaint is a known symptom of depression (i.e., DSM-IV criteria) such as depression, irritability, fatigue, sleep disorder, change in appetite
- Clinical presentation is vague or does not make sense as a physiological process
- Patient seems sad beyond what would be expected from his or her disease
- The physician feels sad talking to the patient
- Multiple chronic illnesses
- Multiple psychosocial stressors
- The physician develops a suspicion for depression

---

setting. Although this cannot be generalized to all patients, it is a possible area for future research. Nonetheless, many patients may be initially reluctant to disclose emotional symptoms and may "test the waters" by discussing physical complaints first.

Given that reluctance to bring up emotional concerns is common, the physician should be alert for subtle cues that suggest underlying depression. The physician should listen for statements that are full of strong feelings, such as strong sadness, loss, or expressions of being trapped or stuck. If the physician starts to develop strong feelings (in this case feeling depressed) while talking with the patient, it may well be a clue to delve into the patient's feelings.

HOW SHOULD THE PHYSICIAN ASSESS FOR DEPRESSION OVER THE TELEPHONE?

The assessment for depression over the telephone requires two basic tasks: creating an environment where the patient feels comfortable disclosing emotions and feelings and asking specific screening questions.

Creating a Comfortable Environment for Disclosure of Emotions and Feelings

Open-ended questions about the patient's functioning and mood may help the patient to open up about his or her feelings. Basic questions for

---

Table 15-2   Empathetic Responses to Strong Emotional Statements

- "It seems you are going through difficult times."
- "That must be very distressing."
- "It seems you're feeling quite desperate."
- "That sounds awful; you must be really sad."
- "It sounds like you are feeling sad about this."

---

opening the discussion include "How are things at home?", "How are things at home with your family?", or "Have you been feeling down lately?"

Communicating concern is important when encouraging a patient to open up. This can be more challenging over the phone; in the office setting a physician can communicate concern with caring and attentive postures and facial expressions. The result may be that the patient chooses not to disclose his or her true feelings. To avoid this, the physician on the phone should be sure to respond to a patient's statement concerning a strong emotion with an empathetic comment (Table 15-2).

Sometimes a physician will get the sense that the patient is experiencing strong emotions, but cannot say exactly what the patient's emotions are. When in doubt, it is good to be curious (12). For example:

> *Doctor:* "I can tell that you have strong feelings about this."
> *Patient:* "Yes, I do."
> *Doctor:* "But I am not sure I understand exactly how you feel. Could you tell me?"

Specific Questions for Screening for Depression

A recent study in a primary care setting found that two questions were predictive for a diagnosis of depression:

1. "During the past month, have you often been bothered by little interest or pleasure in doing things you used to enjoy?"
2. "During the past month, have you often been bothered by feeling down, depressed, or hopeless?"

Table 15-3   DSM-IV Criteria for Major Depression

Five or more of the following nine symptoms should be present for a 2-week period. At least one should be depressed mood or loss of pleasure.

Mnemonic: Dr. SIG: E CAPS ("Doc, Prescribe Energy Capsules")

D—*Depressed mood*	Depressed mood most of the day, nearly every day (reported or observed)
S—*Sleep*	Insomnia or hypersomnia
I—*Interest*	Loss of interest and pleasure
G—*Guilt*	Feelings of guilt, worthlessness, hopelessness
E—*Energy*	Loss of energy, fatigue
C—*Concentration*	Impaired concentration
A—*Appetite*	Decreased or increased appetite or weight
P—*Psychomotor*	Psychomotor agitation or retardation
S—*Suicidality*	Recurrent thoughts of death and suicide

From Marzuk PM. Suicidal behavior and HIV illness. Int Rev Psychiatry. 1991;3:365-71; with permission.

Answering "no" to both of these questions effectively ruled out depression (sensitivity 97%, likelihood ratio for negative result 0.07). Answering "yes" to either did not make the diagnosis of depression certain, but it made it more likely (specificity 57%, likelihood ratio 2.2) (13). If the patient answers yes to either question, a complete screen for depression is indicated (see below), and the physician should help the patient to accept treatment.

If the patient's responses to the above questions or comments suggest possible depression, a more formal depression screen should follow, based on the DSM-IV criteria summarized in Table 15-3.

IS THE PATIENT AT RISK FOR SUICIDE?

Any patient suspected of having new-onset major depression or a recurrence of major depression based on the above screening questions should be screened for suicidality. Some patients will make an indirect statement suggesting suicidality such as, "I've had enough," "I'm a burden," or "It's not worth it," which mandates follow-up with specific questions. Although most patients with suicidal ideation will not go on to commit suicide, there is currently no reliable method for predicting a patient's suicide risk. Some patients will clearly have "active" suicidality, with a clear intent to carry through with a specific plan. Many patients, however, will have "passive" suicidality, which consists of vague thoughts or wishes about dying but no concrete plan to harm themselves. Determining which category the patient fits into can be challenging. There are, however, a number of risk factors that can predict a higher suicide risk (Table 15-4).

The most accurate predictor of completed suicide is a history of attempted suicide. There is some evidence to suggest an increased risk of suicide in patients with cancer, head injury, and peptic ulcer disease (14). Patients with AIDS are 16 to 36 times more likely to die by suicide than persons without AIDS (15).

Once a patient admits to suicidal ideation, it is crucial to inquire about a plan. More violent methods are more likely to lead to "successful"

Table 15-4 Risk Factors for Suicide

1. History of previous suicide attempt (especially with firearms, jumping, or drowning)

2. Male gender

3. Over 45 years of age

4. White race

5. Separated, divorced, or widowed

6. Living alone

7. History of alcohol or drug abuse

8. Chronic medical illness

9. Depression

10. Hopelessness

11. Family history of substance abuse

12. Family history of suicide attempts

13. Psychotic symptoms

14. Patient at the beginning or end of a depressive episode

suicide. In depressed patients, the highest risk for suicide occurs during the period when a patient is beginning to emerge from a depression, when he or she has more energy to carry out a suicidal plan. This is also likely to be the time when the patient may call his or her physician to discuss somatic complaints.

HOW SHOULD THE PHYSICIAN ASK THE PATIENT ABOUT SUICIDE?

A common misconception, especially among primary care physicians, is that asking specific questions about suicide will increase the likelihood of suicide or plant the idea in the patient's mind. Most patients who consider suicide are ambivalent about the act and will feel relieved that the clinician is interested and willing to talk with them about their ideas and plans (16).

Patients who call may not readily volunteer the information that they are suicidal. They may not want to make the physician uncomfortable, and they may sense a physician's uneasiness in talking about depression or suicide. Unfortunately, patients who are considering suicide may feel a strong desire to talk about it but will tend to wait for the physician to bring up the topic. If suicidality is not brought up at all, the patient may feel even more depressed and may conclude, "No one wants to hear about this." A gradual and respectful approach to raising and discussing suicidality is summarized in Table 15-5.

Table 15-5   Approach to Identifying Suicidality

1. "Sometimes when people feel [*down/blue/sad/depressed*], they think life is not worth living any longer. Has this thought crossed your mind lately?"

2. "Have you ever thought about [*harming/hurting/killing*] yourself?"

3. "Have you ever tried to [*harm/hurt/kill*] yourself?"

4. "When was the last time you [*thought about/tried*] this?"

5. "How did you change your mind?"

6. "Are you considering suicide right now?"

Table 15-6   Supportive Statements for the Suicidal Patient

1. "You and I can work this out together."
2. "I appreciate your telling me about your distress. I am here to help you."
3. "I know how to help you and I would like to . . . ."

If the patient is considering suicide, the following questions should be asked:

1. "Do you have a plan for hurting yourself?"
2. "Do you have the means to commit suicide?"
3. "Do you have a gun at home?"

Based on the answers to these questions and the presence or absence of risk factors (Table 15-4), the physician can begin to get a sense of the patient's risk of committing suicide. However, assessing suicidality requires physicians to use their judgement. Answering "no" to these questions may predict a lower risk of death from suicide, but it is by no means a fail-safe method of ruling out suicide risk.

If the patient is currently suicidal, whether active or passive, the physician should offer help and give hope to the patient by making supportive statements (see Table 15-6). An empathetic comment can communicate understanding (17), which may encourage the patient to get help.

HOW SHOULD A DEPRESSED AND/OR SUICIDAL PATIENT BE MANAGED?

Patients Who Are Actively Suicidal

The physician must try to motivate the actively suicidal patient to go immediately to an emergency department or crisis center. If the patient refuses, the physician can initiate involuntary commitment (see below), but to maximize the patient's long-term treatment success, voluntary action is clearly preferable.

Introduce the subject carefully with an offer to allow the patient to be active in seeking care: "I am very worried about your safety. We need to decide now if you want to make a commitment to getting help today." Be very supportive during this phase of the negotiation in order to increase compliance with the recommendation. It is vital for the physician to obtain the patient's full name, address, and phone number before completing the call.

If the Patient Agrees to Get Help Immediately

What to Tell the Patient
- *Give direct and specific instructions:* "You need to go to [the nearest emergency department or crisis center] right now. I will call ahead, so there will be a doctor waiting for you to help you deal with your feelings and thoughts."
- *Inquire about feasibility:* "Do you feel you can go by yourself to the hospital?" If it seems likely that the patient will not make it on his or her own, summon a responsible family member, an ambulance, or the police to escort the patient to the emergency department.
- *Ensure patient's understanding of plan:* "Do you have any questions about what we've talked about?"
- *Ask about patient's immediate plan of action:* "What will you do when you hang up?"

Call the Emergency Department
Communicate to the emergency department physician the patient's information and clarify who is going to further evaluate the patient.

Call the Emergency Department or the Patient Again Later
Call the emergency department or patient again later to make sure the patient has followed through with the plan.

If the Patient Does Not Agree to Get Help Immediately

Involuntary Commitment
If an actively suicidal patient refuses to go to the emergency department, primary care physicians can activate involuntary psychiatric assessment. States vary in their requirements concerning involuntary assessment. The

police and the consulting psychiatrist can provide guidance for the telephone physician. Depending on the state, the physician can activate involuntary commitment alone or seek concurrence from a second physician (usually the psychiatrist in the treating hospital or emergency department). The police will bring the patient involuntarily to the emergency department or psychiatric center, where a psychiatrist will determine if the patient needs to be admitted or not.

What to Tell the Patient

• *Tell the patient about your concern, your plan, and what to expect:* "I understand you do not wish to go to the hospital right now, but I am very worried that you may try to hurt yourself. I am going to speak with a psychiatrist at the hospital, and if he shares my concern for your safety, we will have an ambulance come to your house to take you to the emergency room."

Call the Emergency Department

Call the emergency department and tell the attending physician the patient's name, your concern about suicidality, and that the patient has refused to go to the emergency department voluntarily. Ask to speak with the psychiatrist on call to clarify what the legal details are for your state regarding paperwork (who needs to sign the forms) and transportation of the patient (ambulance and/or police).

Contact the Police Department

Contact the police department and give the patient's name, address, and telephone number, and inform the police that it is your professional opinion that the patient is dangerous to himself/herself. Clarify with the police how the patient will be transported to the hospital.

Patients Who Are Not Actively Suicidal

If the patient has major depression but is not actively suicidal, the most important task is to facilitate appropriate ongoing care. This can be achieved through effective counseling on the nature and treatment of depression and ensuring a prompt appointment with the patient's primary care physician. As a rule, avoid referring the patient immediately to a psychiatrist. The patient may be more open to discussing the problem with the primary care physician, and treatment can often be initiated without referral to a psychiatrist.

Antidepressants should generally not be prescribed over the telephone, with the exception of a patient with a history of depression who responded well to antidepressants in the past but is now off medication. In this case, the same medication can be restarted at the same dose that was previously effective. If the patient does not recall the dose, one can prescribe the lowest recommended therapeutic dose for that medication. This is not to suggest that the telephone contact be used as a replacement for continuity of care for the depressed patient or for emergency department referral of the suicidal patient. However, restarting medication for a recurrence of depression should be seen as an option that allows the patient to start effective treatment promptly. In the case of recurrence, earlier initiation of treatment correlates with a better response rate (2).

What to Tell the Patient
• *Counsel the patient about the diagnosis, its significance, treatment, and prognosis:* "You have what we call major depression. This is caused by an imbalance of substances in your brain. This is the major cause of your symptoms. The good news is that there is effective treatment for major depression, including medication and counseling. With treatment, you have an excellent chance of feeling much better, like your old self. Without treatment, you may get better, but it will take longer. There is also a possibility that you may get worse."

• *Ensure follow-up with the primary care physician within a few days:* "You should see [primary care physician] within a few day to talk more about how you're feeling and what the options for treatment are."

Consider Restarting Medication
If the depressive episode is a recurrence, consider restarting the medication that was previously successful.

Contact the Primary Care Physician
The telephone physician should communicate with the primary care physician directly—either in writing or on the telephone or both—to

PATIENT WHO CALLS WITH POSSIBLE DEPRESSION

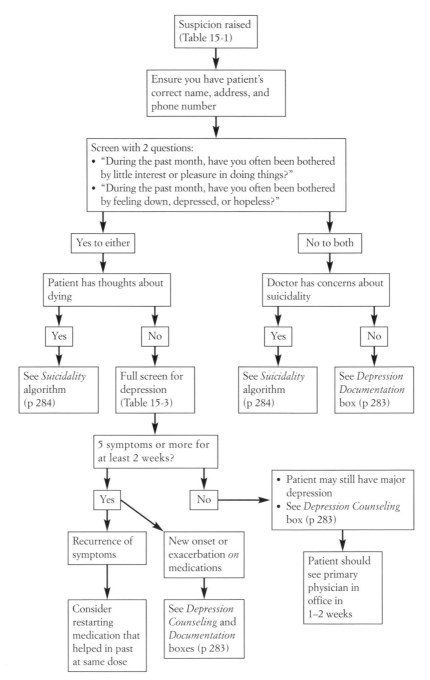

Depression Counseling—Explain the following to the patient:
- Patient has major depression
- It is a chemical disorder causing symptoms
- Treatment often leads to rapid recovery
- When and where to follow-up
- Call back if symptoms worsen

Depression Documentation:
- Why depression was suspected
- Presence/absence of symptoms of depression
- Presence/absence of active suicidality
- What patient was counseled to do
- Evidence of patient's understanding
- Whether patient verbalizes agreement with plan
- Whether emergency department or police was called

---

ensure that all the relevant information is communicated and that a follow-up appointment takes place. Remember that the patient may be embarrassed and may just want to "forget about the whole thing."

Consider Calling Back Later or the Next Day
This will convey to the patient that you take the condition seriously and that he or she should feel comfortable telling the primary physician about his or her depression.

WHAT TO DOCUMENT

Clear documentation about presence of depression and/or suicide is a powerful tool against litigation and for ensuring follow-up of the problem. The physician should document presence or absence of all symptoms of depression and clearly document the presence or absence of suicidal ideation. It will be important to include in the documentation the relevant predictors of suicide from Table 15-4. If suicidal ideation is present, you need to document whether or not the patient is judged to be acutely at risk and why.

Document your recommendations to the patient and whether the patient agrees with the plan. Also document whether the emergency department

Patient Who May Be Suicidal

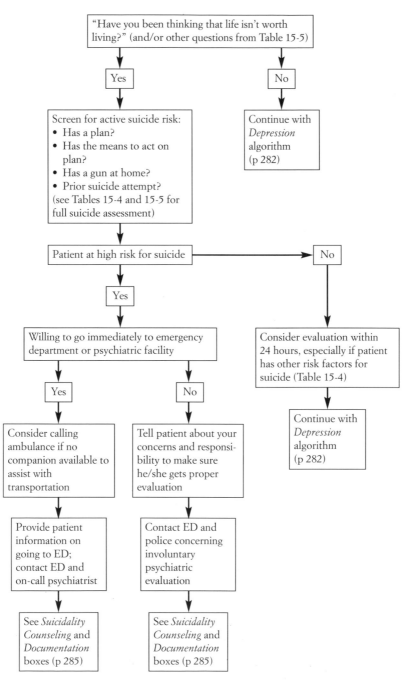

"Have you been thinking that life isn't worth living?" (and/or other questions from Table 15-5)

Yes

No

Screen for active suicide risk:
• Has a plan?
• Has the means to act on plan?
• Has a gun at home?
• Prior suicide attempt?
(see Tables 15-4 and 15-5 for full suicide assessment)

Continue with *Depression* algorithm (p 282)

Patient at high risk for suicide

No

Yes

Willing to go immediately to emergency department or psychiatric facility

Consider evaluation within 24 hours, especially if patient has other risk factors for suicide (Table 15-4)

Yes

No

Consider calling ambulance if no companion available to assist with transportation

Tell patient about your concerns and responsibility to make sure he/she gets proper evaluation

Continue with *Depression* algorithm (p 282)

Provide patient information on going to ED; contact ED and on-call psychiatrist

Contact ED and police concerning involuntary psychiatric evaluation

See *Suicidality Counseling* and *Documentation* boxes (p 285)

See *Suicidality Counseling* and *Documentation* boxes (p 285)

Patient Who May Be Suicidal (cont'd)

Suicidality Counseling—Explain the following to the patient:

If not actively suicidal:
- Patient needs to see primary physician within next few days.
- If symptoms worsen, or patient starts to feel like hurting himself or herself, patient must call back or go to emergency facility.

If actively suicidal/requires emergency evaluation and agrees to go:
- Patient should go to emergency facility *now* (name specific facility).
- Patient will speak with psychiatrist who may recommend treatment and/or admission.
- Encourage patient to be transported to the emergency room by a companion if at all possible. If unavailable, the physician should contact an ambulance service.

If actively suicidal/requires emergency evaluation and refuses to go:
- Explain you are worried that patient may try to harm himself or herself.
- Explain that you are going to have the police take patient to an emergency facility.
- A psychiatrist will evaluate patient to see if patient should stay in hospital for treatment.

Suicidality Documentation:
- Why depression was suspected
- Presence/absence of symptoms of depression
- Presence/absence of active suicidality
- What patient was counseled to do
- Evidence of patient's understanding
- Whether patient verbalizes agreement with plan
- Whether emergency department or police was called

or the police were called, as well as your conversations with the hospital psychiatrist, emergency physician, and primary care physician.

REFERENCES

1. **Alec R.** Emergency psychiatry: suicide. In: Kaplan HI, Sadock BJ, eds. Comprehensive Textbook of Psychiatry, 6th ed. Philadelphia: Williams & Wilkins; 1995.
2. **United States Preventive Services Task Force.** Guide to Clinical Preventive Services, 2nd ed. Baltimore: Williams & Wilkins; 1996.
3. **Wittchen HU, Lieb R, Wunderlich U, et al.** Comorbidity in primary care: presentation and consequences. J Clin Psychiatr. 1999;60(Suppl 7):29-36.
4. **Angst J.** Major depression in 1998: are we providing optimal therapy?. J Clin Psychiatry. 1999;60(Suppl):5-9.
5. **Hirschfeld RM, Keller MB, Panico S, et al.** The national depressive and manic-depressive association consensus statement on the undertreatment of depression. JAMA. 1997;277:333-40.
6. **Tylee A.** Depression in the community: physician and patient perspective. J Clin Psychiatry. 1999;60(Suppl 7):12-16.

7. **Salazar WH.** Management of depression in the outpatient office. Med Clin North Am. 1996;80:431-55.

8. **Hawton K.** Assessment of suicide risk. Br J Psychiatry. 1987;150:145-53.

9. **Peters RM.** After-hours telephone calls to general and subspecialty internists: an observational study. J Gen Intern Med. 1994;9:554-7.

10. **Ustun TB, Von Korff M.** Primary mental health services: access and provisions of care. In: Ustun TB, Sartorius N, eds. Mental Illness in General Health Care: An International Study. Chichester, England: Wiley; 1995:347-60.

11. **Mermelstein HT, Holland JC.** Psychotherapy by telephone. A therapeutic tool for cancer patients. Psychosomatics. 1991;32:407-12.

12. **Platt FW.** Conversation Repair. Case Studies in Doctor-Patient Communication. Boston: Little Brown; 1995.

13. **Whooley MA, Avins AL, Miranda J, et al.** Case-finding instruments for depression. Two questions are as good as many. J Gen Intern Med. 1997;12:439-45.

14. **Mackenzie TB, Popkin MK.** Suicide in the medical patient. Int J Psychiatry Med. 1987;17:3-22.

15. **Marzuk PM.** Suicidal behavior and HIV illness. Int Rev Psychiatry. 1991;3:365-71.

16. **Doyle BB.** Crisis management of the suicidal patient. In: Blumenthal SJ, Kupfer DJ, eds. Suicide Over the Life Cycle: Risk Factors, Assessment, and Treatment of Suicidal Patients. Washington, DC: American Psychiatric Press; 1990:381-423.

17. **Platt FW, Keller VF.** Empathic communication: a teachable and learnable skill. J Gen Intern Med. 1994;9:222-6.

16

ASTHMA

David L. Stevens, MD

KEY POINTS

- Early aggressive home management of asthma exacerbations can lead to decreased emergency department visits and rapid improvement of symptoms.
- A patient's initial response to an inhaled β-2 agonist (poor, partial, or complete/near complete) is the best predictor of higher risk of hospital admission and respiratory failure.
- Patients may be poor judges of their own levels of airway obstruction.
- When available, peak flow meter readings are more helpful in determining the level of airway obstruction than a patient's symptoms or even in-person chest auscultation.
- Patients should be referred for immediate emergency department treatment if there is evidence of severe airway obstruction (i.e., peak flow less than 50% personal best/predicted, severe symptoms, poor ability to speak/walk) despite inhaled β-2 agonist use.
- An acute asthma exacerbation in patients with chronic, poorly controlled asthma, significant cormorbidity, or a history of near-fatal asthma attack carries a higher risk for respiratory failure and death and should prompt strong consideration for emergency department referral for prompt treatment.

Continued

- Rapid onset of an attack, especially less than 3 hours from onset to severe dyspnea, carries a high risk for mortality. These patients should be referred immediately to an emergency department if not responding well to an inhaled β-2 agonist.
- Patients with partial response to an inhaled β-2 agonist (peak flow 50% to 80% of personal best/predicted or persistent symptoms despite significant relief) should be managed as follows:
  1. Continued inhaled β-2 agonist use
  2. A course of an oral corticosteroid, such as prednisone 40 mg daily for 3 days
  3. Patients already taking inhaled corticosteroids should double the number of puffs they are taking to prevent relapse of the exacerbation
  4. Instruction on self-monitoring of peak flow and/or symptoms with specific instructions on when to call back
  5. See primary care provider within 1 to 2 days to evaluate response and need for additional medications, such an inhaled corticosteroids
- Patients with complete or near-complete response should be instructed to self-monitor and call back if worsening; if the patient is already taking an inhaled corticosteroid, the dose should be doubled.

EPIDEMIOLOGY

Asthma is common in adults, affecting approximately 10 million people in the United States. Although hospitalizations are common (470,000 admissions to U.S. hospitals annually), fatalities are less common (approximately 5000 per year, or about one in every 1000 admissions) (1).

Utility of Early Diagnosis

The goal of early diagnosis and treatment is to reduce airway obstruction. If this goal is promptly met, unnecessary emergency department visits and hospitalizations will be prevented, loss of occupational and social function will be limited, patient discomfort will be reduced, and risk of death or respiratory failure may also be reduced. The evidence to support this promotion of early intervention comes largely from studies of patients taught to follow a personalized algorithm for self-management at home (2) and from an intervention that included aggressive pharmacological management, education, and easy telephone access to a physician (3). The guidelines of the National Heart, Lung, and Blood Institute/National Asthma Education and Prevention Program Expert Panel Report (1) for home management is an expert consensus on how patients can be advised about home treatment. The telephone management of asthma discussed in this chapter is based on this document.

Many attacks can be managed at home, especially if treatment is initiated early. A growing body of evidence shows that proper use of medications can prevent attacks (1,2). On the phone, management is greatly facilitated when the patient has been instructed on self-assessment using symptoms (or symptom diaries) and peak flow measurement. Most asthmatics should be trained in the use of Asthma Action Plans (Figure 16-1) in which patients monitor their own symptoms and/or peak flow and begin self-treating mild attacks early but contact a physician for more severe attacks. Evidence shows that early and aggressive home management with an Asthma Action Plan results in fewer emergency department visits and fewer missed days of work (2).

Unfortunately, many patients calling for help with their asthma will not have been trained in the use of an Asthma Action Plan. Others may call after their attempts at self-management have not resulted in adequate improvement. The doctor's role on the phone is to:

- Ascertain quickly if the patient's symptoms indicate an asthma exacerbation.
- Determine if the patient requires immediate treatment in an emergency department.
- Counsel the patient on what management strategies are likely to result in prompt and lasting relief of symptoms if emergency department treatment is not required.

The focus of this chapter, therefore, is to help the physician answer the following central questions:

Asthma Action Plan For _____ Doctor's Name _____ Date _____

Doctor's Phone Number _____ Hospital/Emergency Room Phone Number _____

Take These Long-Term Medications Each Day (include and anti-inflammatory)

Medicine	How much to take	When to take it

GREEN ZONE: Doing Well

■ No cough, wheeze, chest tightness, or shortness of breath during the day or night

■ Can do usual activities

And, if a peak flow meter is used,
Peak flow: more than _____
(80% or more of my best peak flow)

My best peak flow is: _____

| Before exercise | □ □ 2 or □ 4 puffs | 5 to 60 minutes before exercise |

First ⇧

YELLOW ZONE: Asthma is getting worse

■ Cough, wheeze, chest tightness or shortness of breath, or

■ Waking at night due to asthma, or

■ Can do some, but not all, usual activities

—Or:—

Peak flow: _____ to _____
(50–80% of my best peak flow)

Second ⇧

Add: Quick-Relief Medicine—and keep taking your GREEN ZONE medicine

□ _____ □ 2 or □ 4 puffs, every 20 minutes for up to 1 hour
(short-acting beta₂-agonist) □ Nebulizer, once

If your symptoms (and peak flow, if used) return to GREEN ZONE after 1 hour of above treatment:
□ Take the quick-relief medicine every 4 hours for 1 to 2 days.
□ Double the dose of your inhaled steroid _____ (7–10) days.

—Or:—

If your symptoms (and peak flow, if used) do not return to GREEN ZONE after 1 hr of above treatment:
□ Take: _____ □ 2 or □ 4 puffs or □ Nebulizer
(short-acting beta₂-agonist)
□ Add: _____ _____ mg per day for _____ (3–10) days
(oral steroid)
□ Call the doctor □ before/□ within _____ hours after taking the oral steroid.

RED ZONE: Medical Alert!

■ Very short of breath, or

■ Quick-relief medicines have not helped, or

■ Cannot do usual activities, or

■ Symptoms are same or get worse after 24 hours in yellow zone

—Or:—

Peak flow: less than _____ (50% of my best peak flow)

Take this medicine:

□ _____ □ 4 or □ 5 puffs or □ Nebulizer
(short-acting beta₂-agonist)

□ _____ _____ mg.
(oral steroid)

Then call your doctor now. Go to the hospital or call for an ambulance if:
■ You are still in the red zone after 15 minutes AND
■ You have not reached your doctor.

⇧

DANGER SIGNS

■ Trouble walking and talking due to shortness of breath
■ Lips or fingernails are blue

■ Take □ 4 or □ 6 puffs of your quick-relief medicines AND
■ Go to the hospital or call for an ambulance (_____) NOW!

Figure 16-1 Example of an "Asthma Action Plan." From National Heart, Lung, and Blood Institute. Expert Panel Report 2: Guidelines for the diagnosis and management of asthma. National Institutes of Health Publication 97-4051; 1997.

1. Should this patient be treated for an asthma exacerbation?
2. Does this patient require treatment in an emergency department?
3. If emergency treatment is not required, how should the patient be advised about home management?

SHOULD THE PATIENT BE TREATED FOR AN ASTHMA EXACERBATION?

Physicians can provide valuable assistance on the telephone to patients with an established diagnosis of asthma. The first task, however, is to determine if the patient has asthma and whether it is an asthma exacerbation and not another diagnosis. The telephone consultation is not the setting in which to make the initial diagnosis of asthma in a patient with no previous history of asthma because this requires more objective measurements of pulmonary function, preferably spirometry (1). A variety of pulmonary and cardiac conditions can present with dyspnea, cough, chest tightness, and even wheezing. Patients experiencing these symptoms for the first time should be evaluated for other possibly life-threatening conditions and may need emergency department evaluation (see Chapter 4 and Chapter 7).

Similarly, if the presenting symptoms are markedly different from an asthmatic's usual asthma symptoms, it is safest to assume the episode may not be asthma and to consider other diagnoses. Dyspnea that is much more severe than usual, fever, copious sputum, or chest pain (as opposed to tightness) suggest that this episode may require immediate in-person evaluation to rule out other causes of the symptoms, such as pneumonia, myocardial infarction, or congestive heart failure.

Questions useful in the determination of a history of asthma and confirmation that the current episode represents asthma and not another diagnosis are given in Table 16-1. Patients with typical symptoms of dyspnea, wheezing, cough, and chest tightness who have had a diagnosis of asthma made in the past and whose symptoms are usually alleviated with asthma medications probably have asthma. If the current symptoms are similar to the patient's usual symptoms when having an exacerbation, then this is probably an asthma exacerbation. If, on the other hand, the symptoms are qualitatively different, other diagnoses should be considered.

Table 16-1   Questions That May Be Helpful in Confirming That an Episode Is an Asthma Exacerbation

---

* *Has a doctor told you that you have asthma?*

* *Other medical problems? (especially consider conditions that would cause respiratory symptoms)*

* *Medication use:*
  —Are you taking any medications for asthma?
  —How do you feel after taking them?
  —Have you run out of any of your medications?

* *What other symptoms have you been experiencing?*

* *Does this feel like your usual asthma symptoms?*

* *Is anything different from your usual asthma symptoms?*

DOES THE PATIENT REQUIRE TREATMENT IN AN EMERGENCY DEPARTMENT?

Once the physician clarifies that the patient is most likely having an asthma exacerbation, the next step is to determine if emergency department treatment is necessary. Immediate treatment should be considered if the patient has severe airway obstruction despite the use of an inhaled β-2 agonist, if the patient has risk factors for asthma-related death, or if the patient has a rapid onset of attack (especially <3 hours from onset to severe dyspnea) that does not respond well to inhaled β-2 agonist use. Most asthmatics calling with acute asthma exacerbations will have begun to treat themselves with an inhaled β-2 agonist such as albuterol. The patient's response to treatments with inhaled β-2 agonist is an excellent predictor of outcome of the exacerbation (4), and this will guide further decision-making. Based on the response to inhaled β-2 agonist, the physician can classify the exacerbation as poor response, partial response, or complete/near-complete response (1).

Peak flow readings can be of great help in determining whether a patient should go to the emergency department (see Peak Flow section below). When a peak flow meter reading is unavailable, the physician must rely on the patient's history combined with a judicious use of the physician's perceptions of the patient's degree of dyspnea.

Table 16-2   Criteria for Determining Presence of Persistent Airway Obstruction

Criterion 1. Persistent airway obstruction despite 3 doses* of inhaled β-2 agonist as indicated by either of the following:
  • Severe distress with minimal or no improvement with β-2 agonist
  • PEF < 50% personal best or predicted

Criterion 2. Patient has risk factors for asthma-related death (see Table 16-3)

* Dose for acute attack = 2–4 puffs of inhaled β-agonist every 20 minutes.

Patients should be considered for emergency department referral if either of the criteria in Table 16-2 is present (1). The assessment for the presence of these criteria over the phone is discussed below.

Criterion 1: Assessing for Persistent Airway Obstruction

Significant, rapid improvement from inhaled β-2 agonist treatment is the best predictor of maintained clinical improvement (3). Such improvement can be assessed by history, peak flow, and examining the patient for evidence of severe airway obstruction by paying attention to how the patient sounds. The most reliable method for home assessment of the degree of airway obstruction is for the patient to measure his or her peak flow with a peak flow meter and compare it to the baseline or personal best. However, many asthmatics have not been instructed on use of a peak flow meter, so the physician often must rely more heavily on the patient's symptoms. Even when a peak flow is available, a brief history focusing on the acute symptoms is a logical starting point.

Whenever possible, speak directly to the patient. Family members or companions may place the call for the patient, but speaking directly to the asthmatic is important for two reasons:

1. The patient's ability to speak comfortably may reflect the degree of airway obstruction: fluent speech, complete sentences, and prompt responses vs. choppy, delayed one-word responses.
2. The patient can often give a more precise description of the severity of symptoms and past history of asthma than can anyone else.

This does not mean that the impression of a family member/companion should be discounted. On the contrary, the opinion of someone standing right next to the patient may be helpful in assessing the patient's level of distress, especially if the patient seems to be downplaying the severity of the symptoms.

Using the Patient's History to Assess Airway Obstruction

Assessing the degree of airway obstruction using the patient's history can be challenging because many asthmatics are not capable of accurately assessing this themselves. This is especially true of elderly patients and those with a history of near-fatal asthma (1). This is one reason for being especially careful with patients who have risk factors (Table 16-3) for asthma-related death; for example, a patient with known severe disease may think an attack is less severe than it truly is.

Despite these limitations, the patient's history can still provide valuable information for assessing airway obstruction (Table 16-4). A good starting place for determining the severity of the exacerbation is an open-ended question such as "How is your breathing?" If the patient feels the symptoms are severe despite use of β-2 agonist, emergency department referral is indicated. Another method for assessing airway obstruction is to assess any functional impairment; ask if the patient is having trouble performing simple activities such as walking or climbing stairs. A patient's comfort with lying down may indicate severity of airway obstruction. Comfort with lying down suggests a milder exacerbation, whereas asthmatics with more acute dyspnea will usually sit upright (1).

When the physician suspects the patient is downplaying his or her symptoms, perhaps to avoid having to go to the emergency department, the physician may ask permission to speak to a family member. The family member may have the advantage of knowing the patient's baseline, can see the patient directly, and may have seen how the symptoms have been progressing.

Given that most asthmatics have suffered with their disease for many years and have self-managed themselves on a daily basis at least with an inhaled β-2 agonist, it may be worth asking directly, "What made you decide to call now?" or "Do you think you need to go to the emergency room?" A long-term asthmatic may be the best judge of the severity of the exacerbation. When the patient is well educated about his or her asthma,

Table 16-3 Risk Factors for Death from Asthma

History of Severe Exacerbations
- Past history of sudden severe exacerbations
- Previous intubation for asthma
- Previous admission for asthma to an intensive care unit

Asthma Hospitalizations and Emergency Visits
- ≥ 2 hospitalizations in the past year
- ≥ 3 emergency visits in the past year
- Hospitalization or emergency visit in past month

β₂-Agonist and Oral Steroid Usage
- Use of > 2 canisters per month of short-acting inhaled β₂-agonist
- Current use of oral steroids or recent withdrawal from oral steroids

Complicating Health Problems
- Comorbidity (e.g., cardiovascular diseases or COPD)
- Serious psychiatric disease, including depression or psychosocial problems
- Illicit drug use

Other Factors
- Poor perception of airflow obstruction or its severity
- Sensitivity to *Alternaria* (an outdoor mold)
- Low socioeconomic status and urban residence

From Practical Guide for the Diagnosis and Management of Asthma, National Institutes of Health Publication No. 97-4053, 1997; p. 26.

Table 16-4 Questions That May Be Helpful in Assessing the Severity of Asthma

- *Has the [inhaled β-2 agonist] helped? How much?*

- *Are you having difficulty walking? Climbing steps?*

- *How do you feel when you lie down?*

- *Do you think you need to be treated in an emergency department?*

- *What made you decide to call now?*

and certainly when the physician knows the patient well, these questions may be the quickest way of getting to the bottom line.

Important History Beyond the Acute Symptoms

Beyond the assessment of the severity, other historical information can be valuable in determining the need for emergency department treatment.

This information falls into two categories: risk factors for respiratory failure and triggers for the current exacerbation (most commonly upper respiratory infection, running out of medications, environmental/allergic trigger, and stress). Although determining the trigger of an exacerbation may not help acute management, it may help in avoiding future exacerbations. On the other hand, determining whether risk factors for death or respiratory failure are present may lead the physician to decide if a patient requires emergency department treatment.

Peak Flow Measurement

When possible, patients should measure their own peak flow. Accurate peak flow meter readings can be extremely helpful in determining the need for emergency department referral. The physician should always inquire as to whether the patient possesses and has received training in use of a peak flow meter, and if so, what his or her personal best reading is. A peak flow that is less than 50% of a patient's personal best despite use of inhaled β-2 agonist indicates severe, refractory airway obstruction requiring immediate emergency department evaluation. When the patient does not know his or her personal best, the NHLBI guidelines (1) recommend using an estimate: the "predicted" peak flow based on the patient's age and height. Charts of predicted peak flows based on age and height are available from specific manufacturers; rough averages for adults are 300-400 (women) and 450-600 (men). Taller and younger patients will be at the higher end of these ranges. However, estimating the patient's best peak flow, even using the tables, can be flawed because some severe asthmatics may never actually approach their predicted peak flow. A man whose personal best is 350 but whose "predicted" peak flow based on his height and weight is 600 may drop below 50% with even a mild exacerbation. Nonetheless, when the personal best is not known, the predicted peak flow is a conservative estimate, allowing the physician to be reassured if the patient's peak flow approaches the predicted peak flow.

"Physical Exam"

Over the telephone, physicians may perceive indications of the patient's degree of airway obstruction by listening not just to the content of what the patient says but to how the patient sounds. As mentioned above, gauging the patient's breathlessness by his or her ability to speak in complete sentences without stopping to take a breath can be helpful. Difficulty speaking after three doses of albuterol suggests continued high-level airway obstruction and should prompt emergency department referral (1,5).

In contrast, the absence or presence of wheezing audible over the phone is not likely to be helpful. Patients with severe airway obstruction may have an air flow rate too low to generate a wheeze loud enough to be heard over the phone. Patients may experience louder wheezing as their obstruction begins to ease. Even in the office with a stethoscope, physicians cannot accurately estimate the degree of airway obstruction (1). The presence of wheezing is primarily useful as an indicator that the dyspnea may be caused by an exacerbation of asthma. Beyond this, characterizing the wheezing may not be useful.

Mental status should be assessed quickly for a number of reasons. First, an impaired mental status might indicate hypercapnea or hypoxia. Mild agitation may be present with any degree of dyspnea. On the other hand, drowsiness or confusion may signal impending respiratory arrest (1). Second, altered mental status might suggest illicit drug use, which is a risk factor for respiratory failure. Third, impaired mental status may limit a patient's ability to self-manage an exacerbation and may render the patient decisionally incapacitated to decline measures such as emergency department referral. When there is a question, it is prudent to verify any suspicions with a family member if one is available; if the answer is still not evident, it may be best to have the patient evaluated in the emergency department.

Criterion 2: Presence of Risk Factors for Respiratory Failure

Additional history beyond the acute symptoms can be valuable in determining the patient's need for emergency department treatment. Retrospective studies have shown correlations between a number of patient factors and fatality or respiratory failure (1). These factors are summarized in Table 16-3. The presence of one or more of these risk factors suggests a history of very severe attacks, chronically poorly controlled asthma, or significant comorbidity. If the physician does not know the patient well, these risk factors should be explored, especially when the patient's symptoms and peak flow do not seem to warrant emergency department referral on their own merits.

IF EMERGENCY DEPARTMENT TREATMENT IS NOT NECESSARY, HOW SHOULD THE PATIENT BE MANAGED AT HOME?

If the patient's response to inhaled β-2 agonist is partial or complete/near-complete, emergency department treatment is usually not necessary. Home management of patients with poor response is also included here with the understanding that some patients will refuse to go to the emergency department. They may nonetheless benefit from aggressive home management.

Home Management of Patients with Poor Response to Inhaled β-2 Agonist

Patients who report minimal or no improvement with three treatments of inhaled β-2 agonist or have peak flows less than 50% of their personal best should go immediately to an emergency department. However, some may refuse to go to an emergency department despite physician instruction. In this case, the patient should be treated similarly to those with partial responses to inhaled β-2 agonist, with the addition of much closer monitoring by the physician. The patient should start oral corticosteroids right away and come in to the office to be examined as soon as possible. If the patient is unable to come into the office right away, the physician should try to speak on the phone with the patient every few hours to monitor the patient and provide further assistance. Whenever possible, the patient should stay with a family member or companion who can help the patient if they develop worsening respiratory distress. If the patient has been taking inhaled corticosteroids, the dose (number of puffs) should be doubled (1). If the patient has not been taking an inhaled corticosteroid, the physician may consider prescribing one.

Home Management of Patients with Partial Response to Inhaled β-2 Agonist

Patients who note a definite improvement but still have bothersome symptoms and/or peak flows between 50% and 80% of their personal best/predicted should begin a course of oral corticosteroids and be instructed on self-monitoring and follow-up within 1 to 2 days. The physician should also consider high-dose inhaled corticosteroids.

Oral Corticosteroids

The physiological explanation for oral corticosteroid use is that the inhaled β-2 agonist may reverse much of the bronchospasm, but the underlying inflammation persists. When the β-2 agonist begins to wear off in a few hours, the inflammation will lead to more bronchospasm, and the symptoms will return. Very frequent use of β-2 agonist may lead to tachyphylaxis and an asthma attack that becomes much more difficult to control.

A prospective ED study of patients found that a course of prednisone resulted in a far lower rate of recurrence of symptoms and return ED visits than those taking placebo (6).

The optimal duration of treatment with systemic corticosteroids is not clear. There is evidence that prednisone 40 mg daily for only 3 days will decrease relapse. The NHLBI guidelines recommend 40 to 60 mg daily for 3 to 10 days (1). A reasonable approach is to give longer courses to patients with more severe attacks, to those with worse asthma at baseline, and to those with risk factors for poor outcome. More complicated tapering regimens are not necessary with such short courses, especially if the patient is also taking inhaled corticosteroids (1).

Inhaled Corticosteroids

There is emerging evidence that inhaled corticosteroids are also useful for reducing relapse of acute exacerbations. They are not likely to help rapidly relieve the acute airway obstruction, but they do appear to reduce the rate at which obstruction returns once it has been relieved (7). The NHLBI guidelines have long recommended that patients already taking inhaled corticosteroids double their dose (number of puffs) during an exacerbation (1). More recently, evidence has emerged to suggest extending the recommendation of high-dose inhaled corticosteroids to patients not already taking them.

A recent study (7) examined the effect of prescribing inhaled corticosteroids (in addition to an oral steroid course) in patients not previously taking inhaled corticosteroids with an acute asthma exacerbation. Patients discharged from an emergency department were instructed to take high doses of the inhaled corticosteroid budenoside (800 mcg twice daily) for 3 weeks. The treatment group (oral and inhaled corticosteroids) had a significantly lower relapse rate of 13% when compared to the control group (oral corticosteroids only), which had a relapse rate of 25%.

The question of whether inhaled corticosteroids should be recommended in telephone management is difficult to answer. Optimally, any patient with a moderate or severe exacerbation should be seen in person, at least within a day or two (sooner for severe exacerbations). Delaying the initiation of inhaled corticosteroids a day or two may not reduce their efficacy, especially if the patient is also taking oral corticosteroids. Moreover, starting a new inhaler during an acute exacerbation may lead to confusion between the different inhalers and unreasonable expectations that the inhaled corticosteroids will have immediate effects like those of inhaled β-2 agonists. Nonetheless, physicians may decide that an individual patient would benefit from starting inhaled corticosteroids before an office visit. Table 16-5 lists comparable high doses for inhaled corticosteroids (1).

Calling Back

Patients should be instructed to call back if symptoms worsen or stop responding to β-2 agonist. If this occurs, they should be seen immediately in the physician's office or in the emergency department.

Table 16-5   High Doses of Inhaled Steroids for Preventing Relapse of Acute Exacerbation

Beclomethasone (Beclovent, Vanceril)		**Fluticasone Propionate (Flovent)**	
42 mcg/puff	10 puffs bid	110 mcg/puff	4 puffs bid
84 mcg/puff	5 puffs bid	220 mcg/puff	2 puffs bid
Budesonide (Pulmicort Turbuhaler)		**Triamcinoline Acetonide (Azmacort)**	
200 mcg/d	4 puffs bid	100 mcg/puff	10 puffs bid
Flunisolide (Aerobid)			
250 mcg/puff	4 puffs bid		

Office Visit Within a Few Days

Patients with a partial response to an inhaled β-2 agonist should probably be evaluated in person by the patient's primary care physician in an office setting within 1 to 2 days. The purpose of this is to confirm response to treatment, review self-management strategies, and adjust the chronic medication regimen if needed.

Home Management of Patients with Complete or Near-Complete Resolution of Symptoms

Medications

Patients who are feeling close to normal and have peak flows that have improved to greater than 80% of their personal best or predicted do not require oral corticosteroids (1). If the patient is taking inhaled corticosteroids, he or she may double the number of puffs for a 7-day to 10-day period (1). If the patient is not on inhaled steroids, it is probably not helpful to begin one now. Although there are no data for such mild exacerbations, the relapse rate is probably low, so adding inhaled corticosteroids may not provide much benefit in the short term. These patients may take their inhaled β-2 agonist every 3 to 4 hours for the next 1 or 2 days.

Calling Back/Seeing the Patient in Person

Patients should be instructed to call back if symptoms worsen or if they stop responding to β-2 agonist. If they have had no other recent asthma attacks, they can probably wait until their next scheduled visit with their primary care physician.

ARE ANY OTHER TREATMENTS EFFECTIVE FOR ACUTE ASTHMA?

Other treatments have traditionally been used for acute asthma exacerbations, but they are no longer considered to be effective (Table 16-6) (1).

Table 16-6   Treatments Not Recommended for Adults with Acute Asthma Exacerbation

---

- Theophylline (effective in *chronic* use, not *acute*)
- Antibiotics
- Aggressive hydration

WHAT TO TELL THE PATIENT

When Immediate Emergency Department Treatment Is Necessary (Poor Response to Inhaled β-2 Agonist)

- *Go to the emergency department immediately:* "It sounds like your asthma is pretty bad now and isn't getting better with your inhalers. I'm concerned that it could get worse and become life threatening. You should go to the emergency department right now. There's treatment you can get there that will help your breathing."
- *Consider calling 911:* If the patient is alone, is worsening quickly, is extremely uncomfortable, or has significant comorbidity, emergency department transport via ambulance should be considered. The physician should consider placing the call if the patient is alone.

When Immediate Emergency Department Treatment Is Not Necessary (Partial Response to Inhaled β-2 Agonist)

- *Emergency in-person evaluation is not necessary:* "It sounds like you've started to improve with your inhalers. I don't think you need to come in or go to the emergency room at this time."
- *Continue inhaled β-2 agonist treatment:* "I recommend that you continue taking your albuterol [or other inhaled β-2 agonist] two puffs every 20 minutes for the next hour, then every hour or two as needed."
- *Start oral corticosteroids:* "I recommend you start taking prednisone for a few days. This will help your attack clear up faster and keep it from getting worse. I will call in a prescription for you. I'd like you to take 40 mg as soon as you can and then once a day for a total of 3 days." Other dosing schedules may be recommended; see discussion above.

- *If patient is taking inhaled corticosteroids, double the dose for 7 to 10 days:* "I recommend that you start taking more puffs of your [name of inhaled corticosteroid] for the next 10 days. This may keep the attack from coming back."
- *Patient should self-monitor and call back if attack worsens:* "If you start to feel worse, if the [inhaled β-2 agonist] stops working, or if anything happens that you're worried about, give me a call back right away."
- *If patient has peak flow meter:* "Check your peak flow every few hours. Call if it starts to go down."
- *Follow-up visit in 1-2 days:* "I'd like to see you in the office within a day or two just to make sure you are getting better and to discuss the medication with you.
- *Call the patient back:* Consider calling the patient back in a few hours to evaluate the response to treatment. Knowing that he or she is not alone in this may also help the patient feel better and make him or her more likely to call back if the symptoms worsen.

When Immediate Emergency Department Treatment Is Not Necessary (Complete/Near Complete Response to Inhaled β-2 Agonist)

Home management for complete/near complete response is identical to that for partial response to inhaled β-2 agonist, with the exception that oral corticosteroids are not routinely recommended for complete/near complete responses. As with partial responses, it is reasonable to double the dose of inhaled corticosteroids in anyone already taking them. Patients should be instructed to call back if symptoms worsen or if new symptoms develop.

WHAT TO DOCUMENT

Documentation of the call should include:

- Details of the current episode
- Severity of attack/response to inhaled β-2 agonist: whether the patient's response was complete/near complete, partial, or poor

Patient Who Calls with Asthma Exacerbation

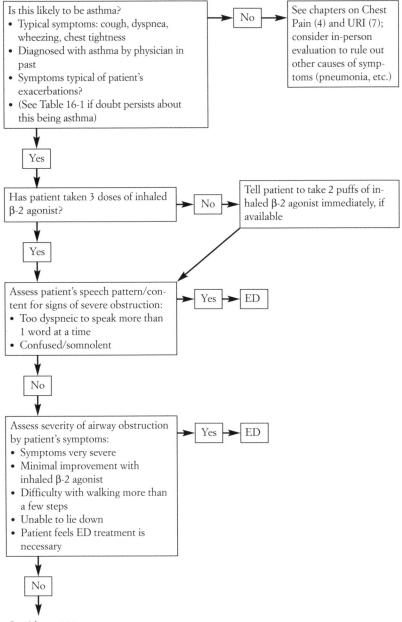

Is this likely to be asthma?
- Typical symptoms: cough, dyspnea, wheezing, chest tightness
- Diagnosed with asthma by physician in past
- Symptoms typical of patient's exacerbations?
- (See Table 16-1 if doubt persists about this being asthma)

→ No → See chapters on Chest Pain (4) and URI (7); consider in-person evaluation to rule out other causes of symptoms (pneumonia, etc.)

Yes ↓

Has patient taken 3 doses of inhaled β-2 agonist?

→ No → Tell patient to take 2 puffs of inhaled β-2 agonist immediately, if available

Yes ↓

Assess patient's speech pattern/content for signs of severe obstruction:
- Too dyspneic to speak more than 1 word at a time
- Confused/somnolent

→ Yes → ED

No ↓

Assess severity of airway obstruction by patient's symptoms:
- Symptoms very severe
- Minimal improvement with inhaled β-2 agonist
- Difficulty with walking more than a few steps
- Unable to lie down
- Patient feels ED treatment is necessary

→ Yes → ED

No ↓

Cont'd on p 305

Cont'd from p 304

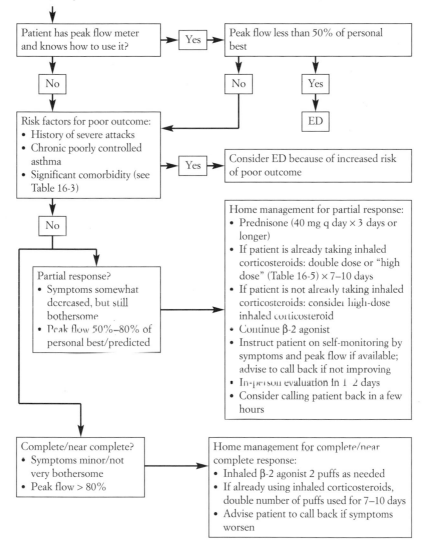

- Patient's exercise tolerance (walking, stairs)
- Chronicity: onset over days, weeks, hours, or minutes
- Medications taken acutely and chronically
- Patient's ability to speak and level of alertness
- Severity of asthma at baseline/recent severe exacerbations/risk factors

- Recommendations made and patient's understanding of the recommendations
- Whether emergency department treatment was recommended; whether patient was told to call 911
- What medications and doses were recommended
- Instructions to patient on how to self-monitor and what symptoms/peak flow the patient was instructed to look for
- When the patient was told to call back

REFERENCES

1. **National Heart, Lung and Blood Institute.** Expert Panel Report 2: Guidelines for the diagnosis and management of asthma. National Institutes of Health Publication 97-4051; Bethesda, MD.
2. **Gibson PG, Coughlan J, Wilson AJ, et al.** The effects of self-management education and regular practitioner review in adults with asthma. The Cochrane Database of Systematic Reviews. 1998;4.
3. **Mayo PH, Richman J, Harris HW.** Results of a program to reduce admissions for adult asthma. Ann Intern Med. 1990;112:864-71.
4. **Rodrigo G, Rodrigo C.** Assessment of the patient with acute asthma in the emergency department. A factor analytic study. Chest. 1993;104:1325-8.
5. **Koury TG, Counselman FL, Huff JS, et al.** Comparison of peak expiratory flow rate with speaking time in emergency department patients presenting with acute exacerbation of asthma. Am J Emerg Med. 1998;16:572-5.
6. **Rowe BH, Spooner CH, Ducharme FM, et al.** The effectiveness of corticosteroids in the treatment of acute exacerbations of asthma: a meta-analysis of their effect on relapse following acute assessment. The Cochrane Database of Systematic Reviews; 1998.
7. **Rowe BH, Bota GW, Fabris L, et al.** Inhaled budesonide in addition to oral corticosteroids to prevent asthma relapse following discharge from the emergency department: a randomized controlled trial. JAMA. 1999;281:2119-26.

17

DIFFICULT PATIENTS

Anna B. Reisman, MD • Mack Lipkin, Jr, MD

Daniel H. Pomerantz, MD

There are certain patients we hate to see on our schedule or who leave us feeling angry or upset after their calls or visits. These patients get labeled "difficult," "obnoxious," "hateful," and worse. The medical literature abounds with discussion of them (1-3). To a great extent, however, which patients are "difficult" depends on the individual practitioner. Some practitioners hate dealing with dependent clingers, others with independent consumerists. Some cringe when they see somatizers on their schedules, others when they see noncompliant alcoholics. Practitioners need to know which types of patients bother them most and either learn specific techniques to handle them better or refer them elsewhere.

A subset of these patients is difficult on the telephone. It includes standard telephone harassers (patients who call regardless of the hour or the significance of their complaint, obscene or seductive callers, abusive callers) and persons who are difficult on the telephone because they are always difficult.

The first part of this chapter focuses on an approach to difficult caller behaviors and how to manage them. The second part focuses on difficult caller types and provides a framework for understanding patients who are always difficult. It focuses on persons with personality disorders as encountered on the telephone and how to address them effectively and efficiently. We end with some aspects of how patients find physicians or their office staff difficult to communicate with on the telephone.

THE TELEPHONE AS A BOUNDARY AND AN ELEMENT OF TIME

Practitioners in their offices are protected from random intrusions by appointment schedules, receptionists, doors, and locks. Curiously, the telephone is a social solvent that dissolves usual barriers. Patients, salespeople, and others reveal their respect for the practitioner's boundaries through their telephone behavior, and reveal their own issues concerning boundaries by when they call and how they request to be spoken to or called back. For example, some demanding patients insist on being spoken to right away, the moment they call. Some call the practitioner's home or pager, at night or on weekends, for minor complaints that more appropriately could be put off until the next day.

Telephone behavior reflects how the caller deals with boundaries between people. A narcissistic patient may not feel separate from the practitioner and may feel free to call when the mood strikes. A borderline or histrionic patient may feel moments of desperate loneliness or emptiness and feel an urgent need for contact with the doctor, using an arbitrary complaint as a "ticket for admission". Thus, when a call comes that seems intrusive and unnecessary, it is useful to view the phenomenon of the call as patient data and to think "Why now?" and "Why in this form?"

In recent years, intrusive calls sometimes reflect the rising costs of care or the inconvenience of coming to the office. It is far easier to call and ask the physician to do something (make a referral, refill a prescription, diagnose a minor complaint) than to come to the office, wait, and pay. With the growth of managed care, patients may feel simultaneously frustrated and entitled and so use the telephone to act out their feelings about the health care system. In parallel, practitioners have set up telephone triage systems ("Please listen carefully to the following options...") that often frustrate and enrage patients who cannot get through, get shunted into blind telephone loops, or spend 10 or 15 minutes to reach answering personnel who are strangers, indifferent, and unhelpful. This aspect of telephone medicine might be called the *difficult telephone system.*

DIFFICULT CALLER BEHAVIORS

Difficult caller behaviors include anger, using the telephone as a way to avoid office visits or paying, using the telephone as a means of contact, intruding at home, using seductive offers or obscene language, rambling, and using the telephone as a means of obtaining narcotics.

Anger

CASE 17-1

The receptionist knocks on Dr. Branford's door during his lunch break to see if he will take a call from a patient, Nora Ryan. Mrs. Ryan is a 43 year-old woman with rheumatoid arthritis whose son is dying of AIDS. Before getting on the telephone, Dr. Branford learns that Mrs. Ryan is furious because she has left Dr. Branford several messages regarding a disability form. Dr. Branford knew of only one call from her that had come in earlier that day; he had planned to get back to her after afternoon office hours.

Possible Reasons for Behavior

Most physicians dislike or dread angry patients. Anger in a patient can be scary or threatening, and may trigger a similar emotion in the physician. An awareness of the likely reasons for a patient's anger can improve the doctor-patient interaction and make the resolution of the conflict easier. Because visual clues from both patient and physician are missing over the telephone, there may be delays in recognizing and addressing anger. The physician would not see the patient's clenched teeth, tightened jaw, and flared nostrils; the patient would not benefit from the physician's nod of empathy or the touch of a caring hand.

Though reasons for anger over the telephone can be straightforward and valid—difficulty getting an appointment, not hearing about test results promptly, frustrations with managed care, they are often less obvious because the anger is displaced from its true object.

Illness

For many patients, anger is a natural response to illness (4). Both newly diagnosed and chronically ill patients often channel their feelings into outbursts of anger, which may be triggered by feelings of helplessness, lack of control, frustration about loss of function, stigma (5), or the shift in how others relate to one that occurs when patients become sick (6).

Displacement

Some patients may have difficulty expressing their anger toward the actual source—an illness and the feelings it generates, or a family member or work colleague—and lose their temper with the physician instead (6).

Personality Disorders

Anger can be associated with some personality disorders. Patients with borderline personality disorder often have free-floating anger, while those with antisocial and narcissistic personality disorders may have angry outbursts when slighted or deprived of something to which they feel entitled (6). These disorders are further described in the section on Difficult Caller Types.

Physician Behavior

Certain behaviors in the physician may provoke anger in the patient. These include being aloof, insensitive, accusatory, or impolite (4,7). When the physician or the system of care humiliates or shames the patient, anger may ensue. A stressed patient who picks up the telephone, hoping for kindness and empathy, but who finds instead a distant or defensive physician, may feel not only let down but enraged.

Positive Aspects

Despite the difficulties of dealing with an angry patient, there are positive aspects as well: anger expressed rather than repressed can clarify a conflict and serve as a trigger to action, possibly improving the doctor-patient relationship (4,7). Anger may simply be a phase of adaptation that will more quickly dissolve not through experiencing defensiveness from the physician but rather through being heard, acknowledged, and accepted. In Mrs. Ryan's case, the patient's anger seems to have come from frustration with not having her calls returned promptly. It is possible that the receptionist took an incomplete message or failed to note the urgency of the call or the date and time. That Mrs. Ryan also suffers from a chronic illness and has an ill son very likely lowered her threshold for anger.

Strategies

Let the Patient Speak Freely
The angry patient needs to vent. The patient may feel that no one—neither office staff nor physician—has taken the time to hear her out. Giving the patient a few minutes to tell the whole story will help exhaust the anger and make her more willing to listen.

Acknowledge the Caller's Feelings
Empathize with the caller's expressed feelings. Statements like "You seem really angry about this" or "It sounds like that experience was really frustrating" can reaffirm to the caller that you have listened and will allow for improved communication (8).

Avoid Becoming Defensive
The reason a patient is angry often has nothing to do with the physician. Even when an outburst feels personal, it can be useful to try to avoid feeling attacked. Becoming defensive may trigger an unending cycle of anger between physician and patient.

Separate the Person from the Problem
Thinking about possible reasons for the patient's anger can make it easier to separate the person from the problem. It is more effective to assess what exactly the issue is and to focus on its resolution, rather than to become angry with the patient.

Apologize When Appropriate
If the patient has legitimate grounds for anger due to the behavior or policies of the physician, office staff, or associated institutions, an apology is

warranted and can be exceedingly helpful. Apology is the form of social convention that serves to heal breaches of decorum, decency, or just plain acceptable behavior. Effective apologies often follow the form shown in Table 17-1.

If the patient's legitimate anger is directed at issues not in the practitioner's control, explicitly stating understanding ("I can really understand why you are angry about that), legitimizing the anger ("Anyone treated that way might get angry"), or even expressing regret ("I am sorry you had to go through that") can help the patient get past her rage.

Use Positive Rather than Negative Statements

Focusing on positives is more helpful than negative statements (Table 17-2). In Case 17-1, for example, blaming the office staff for the problem or echoing Mrs. Ryan's annoyance would be less productive than addressing what can be done to ameliorate the problem. Be direct: "How can I help fix this problem?" The physician may also want to focus on the possibilities for office improvement: "I'm glad you told me that. It will help us improve our service" (4).

Table 17-1   Making an Effective Apology

Statement of the offense:	*I kept you waiting 2 hours.*
Acknowledgment of its importance:	*I know that is wasteful of your time and disrespectful.*
The apology itself:	*I am very sorry to have kept you waiting.*
Intention not to repeat the offense:	*I will do my best not to let this happen again.*
Recompense:	*I will try to make it up to you.*

Table 17-2   Examples of Positive Versus Negative Statements

I disagree.	*vs.*	*I understand; let's discuss the problem.*
Let's compromise.	*vs.*	*Let's find a solution by working together.*
It's our policy.	*vs.*	*This is why it's our policy.*
You have to . . .	*vs.*	*Here's something you may wish to try.*

Modified from Katz HP. Telephone Medicine Triage and Training: A Handbook for Primary Health Care Professionals. Philadelphia: FA Davis; 1990.

Follow-up

Make sure there is a clear plan of action in place to fix the problem. Agreeing on a plan to speak again on the telephone within a designated time period will assure the patient that the problem was important and will be addressed.

RESOLUTION OF CASE 17-1—*Dr. Branford lets Mrs. Ryan vent her anger for a few moments. He starts to explain that he only received one message from her, then changes his mind. Instead, he apologizes. "Mrs. Ryan! Mrs. Ryan!" he says. "You sound so upset, and rightly so. I can hear how frustrating this has been for you. I'm really sorry. Please tell me what I can do to help." Mrs. Ryan takes a surprised breath, pauses for a moment, then asks Dr. Branford if he could have the form ready by the end of the day. Dr. Branford is happy to comply.*

Using the Telephone as a Replacement for the Office Visit

CASE 17-2

Nettie Dunbar, a 50-year-old woman with severe chronic obstructive pulmonary disease, calls Dr. Branford three or four times a month. Over a period of 2 years, Dr. Branford has seen her only twice in the office, which is up three stairs and down a long corridor. Both times she was very impatient with the brief wait, and a third time she stormed out of the office before being seen. Her usual symptoms are shortness of breath and cough, which respond partially to home nebulizer treatments. On the telephone, Dr. Branford usually reviews Mrs. Dunbar's use of her inhalers and oxygen and recommends that she come in for in-person evaluation. Mrs. Dunbar frequently asks about her prognosis as well as her weight loss. Her last pulmonary function tests were more than 3 years ago, and she has missed her appointments for new pulmonary function tests several times. Dr. Branford has explained that he needs more information to address her questions—information that can only be obtained through an office visit and appropriate tests. Nonetheless, Mrs. Dunbar fails to come in.

Possible Reasons for Behavior

There are a number of reasons why some patients use the telephone as a substitute for in-person care. Some patients may have difficulty arranging transportation to the office; others may see the telephone as a cost-saving measure; others, like Mrs. Dunbar, may be impatient with the usual wait at the office. Some, especially cancer patients, may be embarrassed about their appearance (9). Patients like Mrs. Dunbar may find the physical effort of getting into the office taxing, exhausting, and humiliating. Other patients may have a personality disorder, such as narcissism, that gives them the belief that their problems are the most important thing to the physician, who should abide by whatever rules the patient sets.

Strategies

Identify Reason for Patient Reluctance to Have Office Visit
The solution may be apparent once the underlying reason is discovered. A patient unable to afford an office visit may be referred to see a social worker, offered a sliding-scale fee schedule, or, in appropriate cases, told that she can be seen free of charge. A patient unable to arrange transportation can often get helpful information from a social worker. An impatient patient might be told that the office staff will be alerted to his arrival and that the physician will make all efforts possible to see the patient on time. A patient needing assistance getting into the office can call from the lobby, and a member of the office staff can come help her with stairs or other impediments. A patient embarrassed about her appearance can be reassured that such a feeling is normal.

Explain That Telephone is No Substitute for Being Seen
Some patients view the telephone as a complete substitute for the in-person visit. Such patients need to learn that the telephone should be used as an adjunct, not as a replacement. Explain that it is often not safe to change doses or medicines without monitoring physical and laboratory responses. Describe the importance of the physical examination for the specific situation, if appropriate. If home visits are warranted or preferable, they can usually be arranged, although patients may also resist these, especially if appearances matter and the home is difficult to keep clean or make attractive.

Patients with probable personality disorders should be managed in accordance with the section on Difficult Caller Types.

RESOLUTION OF CASE 17-2—*Dr. Branford recognizes that Mrs. Dunbar has a narcissistic personality disorder as well as difficulty getting to the office. He tells Mrs. Dunbar that he understands the importance of her problem and is very concerned about her, adding that he wants to work together to help her. He promises that he will be able to devote complete time and attention to her during the office visit. Dr. Branford acknowledges the difficulty of Mrs. Dunbar's getting to the office and offers to speak with a social worker to help arrange for transportation. It is arranged that Mrs. Dunbar's transporter will call up to the office and that the nurse will come out to help her up the stairs and down the corridor. Mrs. Dunbar is silent for a moment, then agrees to come in the following day, as long as someone picks her up and she is seen on time.*

Calling Frequently for Seemingly Trivial Problems

CASE 17-3

Alvin Wallace is a 56-year-old man with hypertension, atrial fibrillation, and anxiety, for which he takes hydrochlorothiazide, diltiazem, Coumadin, and paroxetine. He is divorced and lives alone. Mr. Wallace is very concerned about taking any medication that might interact with the Coumadin, so he calls Dr. Branford at least twice a week for a variety of reasons: the onset of a cold, a pain in the knee, an ache in the back. He also calls to confirm when he should have his next INR checked and to monitor the status of referrals to various specialists. Today, he calls about a sore neck.

Possible Reasons for Behavior

Patients who call their physicians frequently for minor problems, in addition to coming in for regular office visits, have been described as "high-utilizers." Several studies have found an association between higher levels of utilization and psychological distress in a significant proportion of patients (10-14). Though calls from high-utilizers tend to be more psychological and less organic in nature, patients usually describe physical complaints (15).

Many high-utilizers with psychosocial stress and physical symptoms have psychiatric or personality disorders. Such disorders are generally underdiagnosed in primary care. These include, but are not limited to, somatization, hypochondriasis, anxiety, depression, malingering, chronic pain, and personality disorders. In Case 17-3 Mr. Wallace manifests aspects of anxiety, dependence, and hypochondriasis. What all of these disorders have in common, in the context of the high-utilizer, is the tendency to somatize, or to manifest psychosocial distress through physical symptoms.

Somatizing Patient
The somatizing patient expresses emotional distress through bodily symptoms. Frequent calls may occur for symptoms. Somatized symptoms may make no anatomic or physiologic sense, or may be modeled on previous illnesses of the patient or persons close to him (16).

Dependent Patient
The dependent patient makes frequent calls for trivial reasons because he fears rejection by the physician (3). He uses his symptoms as a means of asking a powerful figure—the physician—to make decisions for him (see section on Dependent Patients).

Hypochondriacal Patient
The hypochondriacal patient calls frequently regarding preoccupations with fear of a serious disease. She may be hoping for either further testing or reassurance.

Chronic Pain Patient
The chronic pain patient makes frequent calls to report persistent pain unrelieved by current treatment regimens. Often, psychological factors can be found that are associated with the onset and maintenance of the pain (16).

Depressed or Anxious Patient
The depressed or anxious patient often expresses his feelings through bodily symptoms. An approach to this type of patient is described in Chapter 15.

Malingering Patient

The malingering patient manufactures symptoms for personal gain. Often, such patients are sociopaths or drug addicts (discussed later in this chapter).

Patient Whose Concerns Have Not Been Addressed

Another explanation of repetitive calls for the same or similar complaints may be that the underlying concern has not been addressed. In Case 17-3, for example, Mr. Wallace may be worried that he is in danger due to his anticoagulation; his physician needs to help him understand how he can live safely while anticoagulated.

Strategies

In approaching patients who call frequently with seemingly trivial problems, it is useful to distinguish between those who may be malingering (see section on Patient Who Seeks Prescription Drugs), those with coexistent anxiety or depression (see Chapter 15), those with an unaddressed underlying concern expressed indirectly, and those who are somatizing. What follows are strategies that are often effective with patients who have a tendency to somatize.

Validate Patient's Distress

Take the patient's symptoms seriously. Perform a careful history and direct the patient in a limited self-directed physical examination, if indicated. The symptoms, no matter how trivial, are important to the patient.

Avoid Ascribing the Symptoms to Psychological Distress

Physicians often ascribe vague symptoms to emotional problems. Patients tend to feel rejected or angered by this. Telling the patient that stress may be worsening the physical complaint, rather than causing it, is generally more acceptable to the patient.

Schedule Regular Telephone Calls or Office Visits

A goal of regularly scheduled telephone calls or regular office visits is appropriate for patients who telephone frequently with seemingly trivial problems. This will help the patient see that he need not have a symptom to maintain contact with the physician. For a physician on a weekend call who is covering such a patient, telling the patient that he will call again during the weekend may be helpful.

Avoid Symptomatic Medication

Prescribing symptomatic medication may exacerbate the situation. The more often a medication is given to treat a symptom, the more the patient will come to expect a prescription for every symptom. Suggest benign treatments, such as vitamins, lotions, and baths, when appropriate (17,18).

Communicate with Physicians Who Share Call

Usually, frequent callers become known to the coverage group. The primary physician can help both the patient and her colleagues by making the call physician aware of the preferred mode of working with the patient. This will minimize ER usage, weekend doctor shopping, and the collection of vast amounts of symptomatic medications.

RESOLUTION OF CASE 17-3—*After ascertaining that the sore neck is most likely due to a strained muscle, Dr. Isaacs expresses her concern for Mr. Wallace's pain. She learns that Mr. Wallace is very stressed about some lingering problems with his ex-wife and tells him that this is likely worsening his pain. Dr. Isaacs suggests the following plan to Mr. Wallace: that he compose a list of ways to deal with his ex-wife before the next appointment, that he stand in a warm shower for 10 min two or three times a day applying the shower needles to the painful area, and that he call her back one week later, even if the pain has resolved. Mr. Wallace agrees with the plan. Dr. Isaacs makes a note to herself to speak about her plan for Mr. Wallace with her colleague who is on call the next weekend.*

Calling Physician at Home Without Permission or About Nonserious Problems

CASE 17-4

Dr. Isaacs and her family are eating dinner when the telephone rings. She lets the answering machine pick up the call. It is Herbert Lawson, one of her patients with diabetes, who apologizes for calling her at home. Dr. Isaacs picks up the telephone. Mr. Lawson explains that his glucometer has been broken for two days and the company has not called him back. He knows that Dr. Isaacs wants him to check his blood sugars every day, and he is not sure what to do. Dr. Isaacs says she will call him back in half an hour.

Soon the telephone rings again. This time it is Joan Burke, a patient who calls her at the office frequently and who has called her at home before. "It's an emergency, Dr. Isaacs," says Mrs. Burke. "My nose has been running all day, and it's so sore I can't take it!"

Possible Reasons for Behavior

Although some physicians will give their home telephone number to a select group of patients, most physicians prefer not to receive direct patient calls at home. Patients who break this barrier and telephone their physician at home usually have a dependent or narcissistic personality style, have experienced great impatience or frustration with the telephone system of the practice, have a feeling of entitlement, or view the relationship with the physician as one of friendship rather than a professional one.

Dependent Patients

The range of illness severity among these patients may be wide, but most have a "self-perception of bottomless need and perception of physician as inexhaustible" (2). Dependent patients tend to defeat every effort to be helped, coming up with new symptoms once one is successfully treated. Reasons for the behavior of dependent patients are discussed in more detail in the section on Difficult Caller Types.

Impatient/Frustrated Patients

Some patients try to reach a physician at home out of frustration. The frustration may be caused by the answering service, an inability to reach the physician through regular means, waiting time at the office, or unreturned messages.

Narcissistic Patients

Narcissistic patients have a feeling of entitlement with their physician. They feel that their well-being is of paramount importance and see no barriers in pursuing the means by which they think they can achieve this, including calling the physician at home.

Patient Who Sees Doctor as Friend Rather Than as Professional

Some patients feel that the physician is a friend rather than a professional, and that this makes calling him or her at home appropriate (see also the section on Seductive Language).

Strategies

Dependent Patient

Inform the patient that there are limits to your time and energy. Ask the patient straightforwardly not to call except during office hours and in emergencies (2). Explain that this is so that you may have the time you need to rest and prepare to be available for the patient and others with equally important concerns.

Another strategy is to view the dependency as a symptom. The patient may be searching for human contact, or perhaps control over the relationship. Sometimes trying to gratify some of the patient's needs may alleviate the problem. Providing telephone numbers for support groups or specific instructions to the dependent patient may be helpful to both you and the patient.

Impatient/Frustrated Patient

First, consider the possibility that the call may be appropriate. In some cases (e.g., where multiple messages have been left because of an emergency and the physician has not called back), it may be appropriate for the patient to try to reach the physician by all means necessary. If this is the case, apologize for not getting back to the patient and address the concern.

If the call is inappropriate, however, address the concern briefly and make a plan to discuss proper telephone use. (In the case involving Mr. Lawson, for example, his frustration is appropriate but calling Dr. Isaacs at home is not.) Assure the patient that you will help him with his concern the following day. During a future telephone call or office visit, review the practice's telephone policy.

Narcissistic Patient

Acknowledge the importance of the patient's problem and, if possible, address it over the telephone. Review the office policy on telephone calls at a future time; if this is already a pattern for this patient, it may be appropriate to discuss the office policy when you return the call.

Patient Who Sees Doctor as Friend Rather Than as Professional

Making a brief statement on the appropriate use of the telephone should suffice. If the office's telephone policy has not been made clear to the patient previously, the policy should be explained. If the patient is angry or upset, it may be more effective to discuss the policy at a later date.

RESOLUTION OF CASE 17-4—*Dr. Isaacs briefly empathizes with Mrs. Burke's sore nose and recommends some Vaseline. She then reminds her about the telephone system that is in place for after-hour calls and tells Mrs. Burke that she must use the system like everybody else. She assures the patient that she will get the proper attention from a nurse or physician on call and asks her to keep a symptom diary to bring to her next appointment.*

After finishing her dinner, Dr. Isaacs calls back Mr. Lawson and tells him that she will have someone look into the problem with the glucometer the next day. She tells him she is glad that he is concerned about the problem but points out that he should call the answering service rather than her directly. She explains that there is always a nurse or doctor who can speak to him who will get the message to her in the morning. Mr. Lawson seems satisfied with the response.

Seductive Language

CASE 17-5

"Doctor, I am so grateful for your care, and I still want to take you out for dinner. When can you go?" Dr. Isaacs closes her eyes and shakes her head when she hears the familiar voice on the telephone. It is Carl Joppa, a 55-year-old widower with diabetes who calls every month "just to check in with my dear friend."

Possible Reasons for Behavior

Seductive behavior can be sexual or maternal or paternal (7). Patients who act seductively generally do so to seek attention. The seductiveness may serve as a way to guarantee that the physician will notice them. Some feel that they will not get such attention unless they give it (4). Often, such patients feel helpless or anxious in the face of unpredictable illness. Seductive behavior can hide feelings of anxiety or lack of self-worth. It can serve to make authority figures seem more humble, especially during an illness where the patient feels unattractive (4).

Strategies

Respond with a Carefully Worded Positive Response
The patient with seductive behavior is trying to get a positive response from the physician. If a positive response is not given, the behavior may continue. Statements such as "You are a colorful person" or "You seem to be doing quite well" (7) will acknowledge to the patient that the physician has noticed the behavior and may, in time, help curb it.

Acknowledge the Patient's Overtures without Responding in Kind
Again, ignoring the behavior will not make it go away. Let the patient know that you consider him or her to be an important person who deserves excellent care.

Set Limits
It is important to set limits with patients and to stick to them. This will help avoid resentment in patients who feel that others might get special treatment from the physician. A statement such as "I really appreciate your invitation for dinner, but it's difficult for me to take good professional care of patients with whom I am socially and personally involved. Because that is such a high priority with me, I simply must decline." (4)

RESOLUTION OF CASE 17-5—*Dr. Isaacs puts the telephone down for a minute and thinks about how to phrase her response. "Mr. Joppa," she says, "I am flattered by your invitation to dinner, but I need to maintain a professional relationship with my patients in order to provide the best possible* *care." Mr. Joppa sighs with disappointment but says he understands and thanks Dr. Isaacs for being straight with him.*

Obscene Language

CASE 17-6

A call is put through one evening to Dr. Isaacs, who is on call. The caller, in between heavy, labored breaths, says his name is Joe Smith. Dr. Isaacs asks Mr. Smith if he is having trouble breathing. "Uh, I wanted to ask you how I can tell if a woman is turned on," he groans.

Possible Reasons for Behavior

The obscene telephone caller derives sexual pleasure from triggering emotion in the caller, whether anger or fear or humiliation (19,20). Obscene callers generally have great difficulties in developing close interpersonal relationships (19), which is why the telephone represents an appealing and safe medium for anonymous communication (21). Obscene callers usually lack insight in seeing that their behavior is problematic.

Although most obscene telephone callers choose their numbers either randomly through the telephone book or by dialing chance numbers (21), the obscene calls that a physician receive may bear more similarity to calls to crisis center hotlines. Obscene callers, hoping to find a female voice at the other end, may try to reach a female physician (21).

One author divided obscene telephone callers into three major types: 1) the caller who informs the listener that he is currently masturbating or who makes obscene propositions; 2) the caller who starts the conversation rationally, develops rapport, then becomes crude and sexually explicit; and 3) the caller whose call is brief but hostile and often describes a violent sexual act (19,22).

Crisis hotline counselors encounter the second type of caller most frequently; this may happen with physicians on call as well. The initial question may be legitimate, while the caller masturbates or breathes heavily and waits for this to be recognized. Such callers are generally repeat callers (21).

Strategies

Hang Up Gently

Because there are no studies specifically addressing how *physicians* can best deal with obscene callers, general advice is given here. Many sources recommend gently hanging up the telephone as a basic strategy (21). One study found that almost a third of obscene callers would not call again if the woman hung up (19). Banging the receiver down, on the other hand, is not recommended; it is a reward to the caller, who may get pleasure from his ability to stir up emotion.

Try To Help the Caller without Engaging Him

Alternatively, the physician may choose to handle the call by trying to help the caller. The challenge is to help the caller without engaging him and fulfilling his desire. One author recommends making a statement such as "It's so sad that you are unable to have sexual feelings unless you are intimidating a woman. Please get professional help." (21)

Avoid Behaviors That May Further Provoke the Caller

Other options—pretending to be hard of hearing, blowing a loud whistle into the telephone, or leaving the telephone off the hook—are not advised. The obscene caller may be inspired to keep trying or may become angry and blow a whistle back.

Contact the Telephone Company or Police

Sometimes, telling the caller that the telephone company and police will be notified is sufficient to stop the calls (21). If the calls occur more frequently or become threatening, however, the physician may want to contact the telephone company to initiate a call trace procedure. Bell Atlantic, for instance, advises its customers to make a police report and, if the calls continue, to call Bell Atlantic's Annoyance Call Bureau; the call will then be investigated if the customer is willing to prosecute once the call is traced.

RESOLUTION OF CASE 17-6—*Dr. Isaacs recognizes that this is an obscene telephone call. She tells Mr. Smith that he has a serious problem and would benefit from therapy. She starts to give him the name of a therapist, but Mr. Smith hangs up. He does not call back.*

Rambling

CASE 17-7

Dr. Isaacs receives a call from Max Rosen, a patient she does not know. Mr. Rosen immediately launches into a lengthy story that starts with his grand-daughter's current visit and moves onto an anecdote about a taxicab driver. Dr. Isaacs tries to interrupt him several times without luck. When she finally does, and asks him why exactly he is calling her, he says he's getting to it and starts to describe what he had for lunch.

Possible Reasons for Behavior

There are many reasons for rambling. When one is rambling, one is not getting to the point. So persons who wish not to get to what is worrying them, or who want to avoid taking responsibility, may ramble. Rambling is a rather direct way of maintaining control of the flow and level of conversation. Rambling serves to distract the listener from the difficult material being salted in and then passed over quickly. It may also indicate a difficulty in concentration, in attention, in language, or in thought.

Strategies

Gently Direct the Flow of Speech Towards a Useful Focus
If the patient continues rambling, gentle interruptions may be needed. It often feels useful to attempt to directly discuss the priorities of the conversations: "Mr. Rosen, we need to focus on a few items so that we can get done what you and I both feel we need to do today. What is your main reason for this telephone call?" This gives the patient some control (of his or her priorities) while asking for some exchange and focus.

RESOLUTION OF CASE 17-7—*Dr. Isaacs manages to interrupt again after Mr. Rosen has a brief coughing fit. In a firm, slow voice she tells him "Mr. Rosen, we need to try to focus on the problem. What is your main reason for this telephone call?" Mr. Rosen pauses for a moment and admits that his heartburn has been worse since his granddaughter's visit and that he is* *worried about it. Dr. Isaacs jumps in again and takes charge of the conversation, focusing on the identified problem.*

Patient Who Seeks Prescription Drugs

CASE 17-8

Dr. Branford is on call over a long holiday weekend. Late on Sunday afternoon he gets a message to call Nancy Gifford, who is complaining of severe headaches. Ms. Gifford reports severe headaches with bilateral temporal throbbing and nausea, which "always gets better with Tylenol #4." She asks Dr. Branford to call in a prescription to her pharmacy and becomes very impatient as he questions her about the details of her headaches. When he suggests that she meet him in the local emergency department for evaluation, she becomes angry and asks Dr. Branford why he is hassling her when Dr. Strauss always calls in her prescription without any problem.

Possible Reasons for Behavior

A small but significant portion of the US population misuses psychoactive drugs. A 1997 survey of the US population aged 12 and older found that 1.2% of the population (2.665 million people) reported nonmedical use of a psychotherapeutic medication in the previous month. Many medications are prone to abuse or diversion, most commonly opioid analgesics, sedative-hypnotics, and stimulants but also muscle relaxants, bronchodilators, antibiotics, and clonidine (used to relieve symptoms of opiate withdrawal).

Patients who seek narcotics or other psychoactive medication over the telephone may have a legitimate reason for the request, either acute pain, such as postsurgical or dental pain, or chronic pain that has been unresponsive to appropriate trials of other modes of treatment. It is well known that many physicians are wary of prescribing narcotics even for cases with a clear indication because of a fear of prescribing something potentially addicting. In addition, many cases of patients requesting narcotics over the telephone, and in person, involve patients who are addicted to the requested substance or who wish to obtain the drug for its street value. Most physicians dread receiving such calls because drug-seeking patients are often effective manipulators. The discussion here will focus on the patient whose request for medication is suspect.

Strategies

Avoid Prescribing Medication Inappropriately and Document the Indications for Treatment

Know the indications, contraindications, and addictive potential of all psychoactive medications you prescribe. One author divided physicians involved in prescription drug abuse into four groups (23):

- *Script Doctors*—a small number of willful misprescribers, whose motive is profit
- *Impaired Doctors*—who prescribe drugs for themselves
- *Duped Doctors*—who acquiesce to patients' demands for inappropriate prescriptions as an expedient way to avoid or limit confrontation
- *Dated Doctors*—who are uninformed or ignorant prescribers

Educate patients about side effects, intended duration of therapy, and withdrawal of therapy. Clear documentation of the indications for treatment is crucial, as is a detailed discussion with the patient about the treatment plan at the initiation of the treatment. Physicians may also contribute to the problem when they prescribe drugs of abuse for patients they do not know without reviewing the medical record or verifying information offered by the patient to support their request for a prescription.

Know When To Be Suspicious

As in office encounters, there are often clues that the problem the patient is seeking to treat is addiction, and not, for instance, headache, back pain, or a kidney stone. Prescription drug abusers may disclose that they are using medication not for its intended purpose but to relieve stress or to help with sleep. Patients with addiction may request a specific medication or formulation, or they may request branded medication, because the generic "doesn't work for me" or "I am allergic to [the generic formulation]".

Table 17-3   Common Ruses Used by Prescription Drug Seekers

I was mugged . . .

I lost my purse . . .

My boyfriend/girlfriend stole my meds . . .

I'm here for a short trip and I forgot to bring my medication . . .

My doctor forgot to write my prescription . . .

Modified from Bigby JA, Parran T. Prescription drug abuse and benzodiazepine abuse. In: Bigby JA, ed. Abuse in General Internal Medicine: A Manual for Faculty. Washington DC: Bureau of Health Professions (HRSA and Office of Treatment Improvement [ADAMHA]); 1993; and Parran T. Prescription drug abuse: a question of balance. Med Clin North Am. 1997;81:967-8.

Branded medications have higher street (resale) value. Addicted patients may try to justify higher doses or more frequent dosing intervals than are usually necessary (although so may patients in pain). There are a number of common excuses for obtaining prescription drugs of which physicians should be aware (Table 17-3). Patients may also initiate encounters at times when the prescriber is under time pressure to decrease the likelihood of a protracted confrontation.

Minimize the Problems of Prescribing Controlled Substances
Ensure that patients who need controlled drugs (patients with malignant pain, severe acute trauma, sickle cell disease) have a supply sufficient for their needs and that they understand how to use them properly. Encourage patients to be proactive in seeking refills during regular office hours to ensure good communication and documentation of prescriptions and medication usage. Document the indication, expected outcome, and duration of treatment in the medical record each time you prescribe medications with potential for abuse.

Obtain Confirmation of the Patient's Story before Prescribing
When unfamiliar patients call to request prescriptions, the physician should obtain confirmation of the story before prescribing, either through chart review if possible or at least by calling the patient's regular pharmacy. Most states require that controlled substances can be prescribed only after a patient with a relationship with the practitioner has been seen and examined. When covering for other doctors, referring to their records may be sufficient.

If, after a telephone call, it seems appropriate and necessary to prescribe a medication with potential for abuse, it is prudent to prescribe the smallest quantity possible to afford the patient relief and to permit the patient to follow up during regular office hours. State regulations frequently limit the quantity that can be prescribed by telephone to a 5-day supply or less.

Learn To Feel Comfortable Saying "No"

Many prescription-seeking patients have had the experience of turning an initial "no" into a "yes" by persevering in their demands (24). Saying "no" to writing a prescription does not mean refusing to offer care to the patient; it means offering the patient care appropriate to the patient's condition. In Case 17-8 treatment for headache would include nonpharmacologic treatment and non-narcotic medication.

Turn the Tables on the Patient

In the face of patient pressure to write an unnecessary prescription, you can turn the interview back to the patient: "I am feeling pushed to write a prescription that is not clinically indicated. Because of this I am really concerned about you, and we need to talk about your use of alcohol (or other substances)."

By turning the tables you reframe the problem as possible substance abuse/misuse and you shift the discomfort to the patient and away from yourself.

RESOLUTION OF CASE 17-8—*Ms. Gifford angrily refuses to discuss her headaches in any further detail and tells Dr. Branford she will not go to the ED. Dr. Branford offers her general advice on other measures to manage headache pain including caffeine, acetaminophen, and NSAIDs. After the weekend, Dr. Branford finds out from her colleague that this patient frequently calls the office requesting prescriptions but almost never follows up with scheduled appointments.*

DIFFICULT CALLER TYPES

As discussed earlier, some patients have personality styles that predictably lead them to difficult interpersonal behavior that extends to telephone behavior. In this section the same physical problem plays out differently as a function of the patient's personality style. The brief vignette is followed by a précis of the underlying nature of the personality disorder and by guidelines for managing the disruptive telephone-related behavior.

"Personality styles" describe stereotypes of behavior that some patients fit. When the fit is good, they are useful because they predict how patients respond and can guide patient interaction. They are believed to be developmentally derived but are also genetically and culturally influenced (3). Personality styles that are maladaptive or disruptive are described in DSM-IV as Axis II disorders, organized as in Table 17-4.

In each of the following examples the patient is a 39-year-old man who calls with abdominal pain, diarrhea, and abdominal bloating that began 2 hours earlier. Although his symptoms are the same in each example, the personality type of this patient is different in each and therefore needs to be handled differently. For convenience, we will call each of these patients "Mr. Hardy."

Table 17-4   Personality Style Groups

Style	Predominant Affect	History of Relationships with Others
Odd/Eccentric		
Paranoid	Guarded, suspicious, flat	Limited, uninvolved,
Schizoid		lacking emotions
Schizotypal		
Dramatic/Emotional/Erratic		
Antisocial	Pronounced, dramatic	Quick, intense beginnings
Borderline		but abrupt, dramatic
Histrionic		endings
Narcissistic		
Anxious/Fearful		
Avoidant	Anxious, fearful	Few relationships, limited
Dependent		involvement
Obsessive-compulsive		
Passive-aggressive		

Modified from Putnam SM, Lipkin M Jr, Lazare A, et al. Personality Styles. In: Lipkin M Jr, Putnam SM, Lazare A, eds. The Medical Interview: Clinical Care, Education, and Research. New York: Springer-Verlag; 1995:251-74.

Histrionic Patients

The histrionic Mr. Hardy calls: "Dr. Knight, I am having the worst pain of my life, I am running to the bathroom every few minutes and this water gushes out of me, it is relentless, and I know something terrible is going to happen, I have been calling for hours, don't you care what happens to your patients?"

Histrionic persons are driven by a need to feel emotions and create attachment. They express themselves dramatically, with more effort centered on their feelings than the factual content of what they are expressing. They initiate and cut off relations quickly. They may be seductive or elicit rapid rescue fantasies (you imagine, well, I can fix this poor person that others have neglected) in their clinicians.

Because they feed us what we want to hear, one should use open ended questions and not lead these patients in questioning them. These patients should be assured of contact and caring. Boundaries must be carefully monitored, because these patients push boundaries of closeness and intimacy but get scared when they get too close.

Recognizing that she is dealing with a histrionic patient, Dr. Knight responds: "Mr. Hardy, I am really glad you called and got me. I can hear your concern, and I want to get this under control expeditiously. To help us both understand what is going on, I have a few focused questions for you about what is happening. Tell me when it started. . ."

Narcissistic Patients

The narcissistic Mr. Hardy calls: "Dr. Knight, this is Mr. Hardy. Every one over there knows me. I really don't want to talk with you, but I need your pre-approval, what a nuisance, so I can call the head gastroenterologist, Dr. Drossman. Don't worry, I know what is wrong with me, it requires the best gastroenterologist in town, so send me the approval form dated today and I will call him immediately. He wouldn't waste my time going over redundant details."

Narcissistic patients regard the world from within a thin shell of arrogance and deeper loneliness, fragility of ego, and inability to understand others.

They feel threatened by illness that undercuts their habitual grandiosity. They expect others to know what they want and what they are thinking. They want the best doctors and the most important specialists and are insulted at the implication that their problem is ordinary or does not need specialist care.

Managing a narcissistic patient requires patience. One must manage one's own ego to allow the patient his self-importance without succumbing to his most flagrant demands. Acknowledge the patient's importance, the significance of his complaints, and the potential value of his demands while proposing to work together for now.

Recognizing that she is dealing with a narcissistic patient, Dr. Knight responds: "Mr. Hardy, of course I know who you are, everyone here has heard of you. I can understand why you want to call Dr. Drossman right away. Unfortunately, the practice rules require that I evaluate you first. Fortunately, I have worked enough with Dr. Drossman that he has trained me in just how to get started. So why don't I begin by taking the same history Dr. Drossman would, then we can decide how to proceed. Tell me when this problem started. . ."

Antisocial Patients

The antisocial Mr. Hardy calls: "Dr. Knight, this is Bobby Hardy. I am Dr. Branford's patient and he knows my story inside out. I have this terrible abdominal pain and it just came back. Probably a kidney stone, you know, my urine is a little reddish, I felt like I passed it, but the pain is still unbelievable. He gives me Demerol but I ran out and the prescription is expired. So I just need you to write me a prescription for some, I'll send over my wife, I couldn't even get up to go to the emergency room. Might as well make it for several days worth so I am covered for the next time too, you know, it saves prescription fees and everything. Oh, I'm also having some diarrhea, so could you write me for some Lomotil, same thing, I ran out the last time I had that."

Antisocial patients have underlying impulsiveness and a lack of social conscience. They attempt to use practitioners to obtain what they desire at the moment, often drugs for use or sale. They may have high-level manipulative skills and like good chess players attempt to put the practitioner into a configuration of issues that induces them to take the easy path of providing what is sought.

Dr. Knight says: "Mr. Hardy, I understand how difficult this must be for you. Given the potential seriousness and the pain you are experiencing, I would not feel safe just prescribing over the telephone. How about you meeting me in the ED so we can properly evaluate this thing. If it is too difficult for you to come by car, you can call an ambulance. Then we can check your urine, do a cystoscopy or ultrasound if need be. . ."

Mr. Hardy interrupts: "Doc, I can't wait, just send the prescription over with my wife..."

Dr. Knight: "Mr. Hardy, I can't do that over the telephone."

Mr. Hardy: "Oh never mind!" He hangs up.

Borderline Patients

*Mr. Hardy calls, "Dr. Knight, I am having the worst night of my life. You have to do something or I will do something drastic. I can't stand this diarrhea. This abdominal swelling, I feel pregnant, the pain is like a delivery, I need relief, I am coming over there now." He hangs up. Dr. Knight calls back using *69. Mr. Hardy answers, "Is that you, Doc? I knew you wouldn't let me down. I'm feeling better but I still need to meet you right away. Where? I'm not going to the ED, I hate how they treat you there, take a number like at the butcher's. You never do that, you are an angel. This pain, what am I going to do?"*

Borderline patients are among the most challenging and provocative callers. They are felt to be developmentally "primitive" personalities given to impulsive acts and moments of "micro-psychosis". They are histrionic, dramatic, given to splitting the world into bad (the ED) and good (the angels of the moment), and initially engender feelings of closeness and rescue and then suddenly leave in anger or rejection. Managing them initially involves creating a relationship that is not prematurely intimate or rescuing but reliable and strong enough to withstand splitting and rejection. Later management includes setting limits, seeing them regularly so their abnormal illness behavior does not get reinforced, and not getting involved in taking sides or going beyond one's good judgment in care.

Dr. Knight says: *"Mr. Hardy, I can understand your distress and want to help in any way I reasonably can. For now, let's see if we can keep you away from the ED and try to get your most difficult symptoms under control. You will start to feel better soon, but not immediately. Here's what I would like you to do. . ."*

Dependent Patients

The dependent Mr. Hardy calls, "Dr. Knight, I don't know what to do. I have diarrhea, my belly is bloated, I am having such pain! I really need you to see me now. I know it is midnight, but I am at my wit's end. Can you help me? Oh, don't hang up, I'm going to put the telephone down while I rush to the bathroom, be back in a few minutes. . ."

This Mr. Hardy is clinging, dependent, and without awareness of his lack of clear boundaries, intrusiveness, or self-absorption. Such patients feel as if they will suck the very milk of human kindness out of us and leave us depleted and exhausted. They expect us to care for them as if they were infants needing diapering. They respond to structuring, attention, statements of caring, and very clear limit setting.

Dr. Knight: *"Mr. Hardy, I can hear how concerned you are. I want you to relax and try not to worry so much, I am going to help you and we will take care of this together. You will start to feel better soon, but not immediately. These are the steps we need to take. . ."*

Schizophrenic Patients

The schizophrenic Mr. Hardy calls: "Dr. Knight, this is me. You know. I think something is inside me and taking over. I feel shitty. Is this telephone safe? (Long pause.) I didn't think so. I'm going to call you from a safe telephone." Mr. Hardy hangs up and calls back in 20 minutes. "It's me again, I think something bad is taking over. I keep shitting it out but it's still in there. I stopped taking my medicine because it kept it all in."

This Mr. Hardy has ideas of reference, loose associations, and paranoia. He needs to be talked to in very concrete, specific, unthreatening terms.

He needs to be managed through prompt contact, restoration of effective drug treatment for his schizophrenia, and such therapeutic contact as is available to him.

Dr. Knight: "Mr. Hardy, we can manage this together. I want to get you into care promptly to get this taken care of. How about meeting me at the ED where we can get you under care that will help you feel better soon?"

PHYSICIAN TELEPHONE BEHAVIOR

Most of the foregoing comments apply as strongly to physician telephone behavior with patients as vice versa. Patients also find anger, rambling, or rudeness unsettling. Patients have a number of other predictable dislikes and taboos concerning physician telephone behavior.

Many patients are frustrated by impenetrable telephone systems that require listening to multiple messages. Such messages are often confusing and lengthy and sometimes do not cover the options most patients have in mind. The situation is made worse when the patient is in distress or panicked, when the patient hears poorly, when the language is not clear to the patient, or when, ultimately, the patient gets no option to speak directly with a human being. Similarly, patients are often annoyed by answering machines that are on when the office is officially open.

A second series of complaints has to do with what happens when a patient actually gets through to a human being. Being put on hold indefinitely or seemingly indefinitely, getting someone on the line and then losing him, or hearing an unhelpful version of "not my department" can be very frustrating for the patient. Not having a chance to talk with someone who can move a concern or complaint along furthers frustration. Similarly, not getting a call back from the doctor or someone who can address the patient's concerns is a frequent complaint, which is also potentially a medicolegal issue.

Patients find the physician's telephone manners significant (see Chapter 2). Does the physician listen? Does she pay attention long enough to hear the real concern? Does she respond respectfully and make a plan, disposition, or referral? Does the physician cut the patient off or hang up peremptorily? Does the physician remember the patient (challenging for those with large, busy practices)? Remote computer access to problem and medication lists may ease the difficulties of physicians covering for someone else whose patients are unfamiliar.

Patients also notice when the physician on the telephone is intoxicated, asleep, or distracted.

CONCLUSION

The first step in managing difficult callers is to recognize one's reaction to the patient. Secondly, the physician should attempt to understand why the patient is being difficult and then use the analyses given in this chapter to guide effective management. At the same time, the physician should reflect on his own telephone behavior and be sure not to be a difficult caller oneself.

REFERENCES

1. **Drossman D.** The problem patient: evaluation and care of medical patients with psychosocial disturbances. Ann Intern Med. 1978;88:366-72.
2. **Groves J.** Taking care of the hateful patient. N Engl J Med. 1978;298:883-7.
3. **Putnam SM, Lipkin M Jr, Lazare A, et al.** Personality styles. In: Lipkin M Jr, Putnam M, Lazare SM, eds. The Medical Interview: Clinical Care, Education, and Research. New York: Springer-Verlag; 1995:251-74.
4. **Lipp M.** Dealing with difficult patients. In: Noble J, ed. Textbook of Primary Care Medicine, 2nd ed. St. Louis: Mosby; 1996:1635-43.
5. **Goffman E.** Stigma: Notes on the Management of Spoiled Identity. New York: Simon & Schuster; 1963.
6. **Barsky A.** Approach to the angry patient. In: Goroll AM, Mulley A, eds. Primary Care Medicine: Office Evaluation and Management of the Adult Patient. Philadelphia: JB Lippincott; 1995:1060-1.
7. **Lipp MR.** Respectful Treatment: The Human Side of Medical Care. Hagerstown, MD: Harper & Row; 1977.
8. **Curtis P, Evens S.** The telephone interview. In: Lipkin M, Putnam S, Lazare A, eds. The Medical Interview: Clinical Care, Education, and Research. New York: Springer-Verlag; 1995:187-95.
9. **Mermelstein HT, Holland JC.** Psychotherapy by telephone: a therapeutic tool for cancer patients. Psychosomatics. 1991;32:407-12.
10. **Katon W, Berg AO, Robins AI, Risse S.** Depression: medical utilization and somatization. West J Med. 1986;144:564-8..
11. **Lefevre F, Reifler D, Lee P, Sbenghe M, et al.** Screening for undetected mental disorders in high utilizers of primary care services. J Gen Intern Med. 1999; 14:425-31.

12. **Karlsson H, Lehtinen V, Joukamaa M.** Psychiatric morbidity among frequent attender patients in primary care. Gen Hosp Psychiatry. 1995;17:19-25.
13. **Karlsson H, Lehtinen V, Joukamaa M.** Are frequent attenders of primary health care distressed? Scand J Health Care. 1990;13:32-8.
14. **Hoeper EW, Nyczi GR, Regier DA.** Diagnosis of mental disorders and increased use of health services in four outpatient settings. Am J Psychiatry. 1980;137:207-10.
15. **Daugird AJ, Spencer DC.** Characteristics of patients who highly utilize telephone medical care in a private practice. J Fam Pract. 1989;29:59-64.
16. **Kaplan C, Lipkin M Jr, Gordon G.** Somatization in primary care: patients with unexplained and vexing medical complaints. J Gen Intern Med. 1988;3:177-90.
17. **Barsky A.** A 37-year-old man with multiple somatic complaints. JAMA. 1997; 278:673-9.
18. **Kaplan C, Lipkin M Jr, Gordon GH.** Somatization in primary care: patients with unexplained and vexing medical complaints. J Gen Intern Med. 1988;3:177-90.
19. **Stones C.** Who's speaking? A tentative profile of the obscene telephone caller. J Soc Behav Personality. 1992;7:639-48.
20. **Karpman B.** The Sexual Offender and His Offenses. New York: Julian Press; 1954.
21. **Matek O.** Obscene phone callers. J Social Work Human Sexuality. 1988;7:113-30.
22. **Leising P.** The negative effects of the obscene telephone caller upon crisis intervention services. Crisis Intervention. 1985;14:84-92.
23. **Devenyi P.** Prescription drug abuse. Can Med Assoc J. 1985;132:242-3.
24. **Parran T.** Prescription drug abuse: a question of balance. Med Clin North Am. 1997;81:967-8.

PART III

....

INCORPORATING TELEPHONE MEDICINE INTO THE WORKPLACE

18

ELECTRONIC ADVANCES

Sary O. Beidas, MD

"640K ought to be enough for anybody."

BILL GATES, 1981

Almost one hundred million Americans are using electronic mail (e-mail), 65 million are using the Internet, and these numbers continue to increase worldwide. Not surprisingly, these relatively new forms of communication are starting to be incorporated into the lives of physicians and patients. Online services, in contrast to other communication tools such as the telephone, shift control to the end-user, in this instance the patient. With Internet access, patients are not locked into their doctor's opinion as their only source of information regarding prescription drugs or diseases; they can educate themselves on such issues via the Internet. Other technological advances are also rapidly changing the way we communicate and live. Like e-mail and the Internet, other evolving communication technologies such as telemedicine and interactive telephony promise to reshape our future.

ELECTRONIC MAIL

E-mail Versus the Telephone

Like the telephone, e-mail delivers information instantaneously. One of the benefits of e-mail is its asynchronous nature in which messages volley back and forth over time. The frustrations of "phone tag" and menu-driven voice mails are eliminated by e-mail. E-mail is easily transported (print or electronically stored) to become a permanent part of the patient's medical record. However, e-mail can be intentionally or unintentionally forwarded to or intercepted by other individuals, a feature that makes it less secure than telephone lines.

E-mail Versus a Letter

E-mail has emerged as a hybrid of the spoken and written word through the use of an electronic device. In most instances, e-mail messages are more succinct, specific, and spontaneous than traditional letters. Like postal mail, e-mail has a unique mailing and return address. It also displays a "postmark" indicating the time it was sent and the time it was received. To lessen the chance of a letter being tampered with, the post office provides the option of a registered letter; similarly, an e-mail user may encrypt his or her e-mail to scramble the message content until it is received by the intended party. Many users of e-mail have adopted the use of acronyms and expressive icons in their messages (e.g., "BTW" for "by the way" and "IMHO" for "in my humble opinion").

How Is an E-mail Message Structured?

An e-mail message is composed of three distinct sections, similar to the office memo. The first section, the "Header," contains the delivery information: the e-mail address of the sender, the e-mail address of the recipient, and the names or e-mail addresses of person(s) receiving copies of the message. The header also includes the "Subject" line, which identifies the focus of the message. The second section, the "Body," contains the text of the message. The final section, the "Signature," identifies the sender and may include contact information (title, address, phone, fax, e-mail address, URL if the sender has a personal Web page) (1) .

Is E-mail an Effective Way for Doctors and Patients to Communicate?

There is a dearth of reports on whether e-mail is an effective method of communication between patients and their health care providers. Balas et al evaluated distance medicine technology and found that computerized communication has benefits in the areas of preventive medicine such as management of degenerative joint disease, cardiac rehabilitation, and diabetes care (2). An excellent example of this powerful new medium is illustrated by Jackler in which one of his patients uses a photo attachment to an e-mail document showing his postoperative wound on the first day after surgery (3).

How Commonly Is E-mail Used Between Doctors and Patients, and What Are Physician Concerns?

An increasing number of physicians are using e-mail to communicate with their patients. In general, there are mixed feelings among physicians on the use of e-mail (4). Neill reported in a small survey that patients felt that e-mail is a useful way of communicating with their health care providers (5). According to surveys done by Cyber Dialogue (6), 48% of online users expressed an interest in using e-mail to communicate with their doctors, but only 3% were doing so. In another recent electronic survey by Healtheon, 33% of surveyed physicians indicated their use of e-mail to communicate with patients (7) .

The survey also showed that physicians have four main concerns about e-mail use for patients:

- Confidentiality and privacy
- Overburdening the workload
- Liability concerns
- Problems of access for patients

Although some physicians may feel that these issues may make the regular use of e-mail too complicated and burdensome, it is useful to keep in mind the initial reaction to other new technologies in medicine. When the stethoscope, for example, was invented by Laennec in 1816, some physicians expressed "suspicion and distrust" and viewed it as "a dangerous instrument." (8). By carefully evaluating new communication technologies and developing rules for using them most effectively, such inventions have the potential to become important additions to the repertoire. Table 18-1 lists some advantages and disadvantages of e-mail in patient-physician communication.

Table 18-1 Advantages and Disadvantages of Using E-Mail in Patient-Physician Communication

Advantages

- Improved access (communication, education, scheduling, billing, other)
- Improved efficiency (rapid, accurate, frees up support staff)
- Asynchronous (patient and physician need not be present at the same time)
- Provides a durable copy for record
- Inexpensive
- Rapid dissemination of information
- Can include links to other sites for education or services
- Eliminates "phone tag"
- Convenient for patient and physician
- Allows use of binary data (images, video, graphics)

Disadvantages

- Requires updating of software to keep encryption effective
- Not suited for urgent communication
- Not suited for communicating some test results or breaking bad news
- Not suited for making a diagnosis
- Message may be lost in cyberspace
- Loss of the in-person doctor-patient relationship
- Not reimbursable
- Lack of physician comfort with new technology
- Potential medicolegal issues
- Requires the development of policies and possibly hiring additional support
- Inequitable access (cultural, language, socioeconomic factors)

Guidelines for Using E-mail for Patient-Physician Communication

Effective use of e-mail in a clinical setting is aided by guidelines for physician-patient communication and medicolegal issues (Tables 18-2 and 18-3) (9). It is important to establish rules for e-mail in one's practice that will maximize its communication potential but also protect the physician from medical liability.

Developing a patient-provider agreement to address the items listed in Tables 18-2 and 18-3 is highly recommended to avoid communication problems. The agreement should include details of how each item is

---

Table 18-2 Communication Guidelines

- Obtain patient's informed consent before using e-mail for correspondence.

- Establish a turnaround time for messages. Instruct patients not to use e-mail for urgent matters.

- Inform patients about privacy issues. Patients should know if messages are viewed by anyone other than the physician during addressee's usual business hours or during vacation or illness and that the message will be included as part of the medical record.

- Establish types of transactions (such as prescription refill, appointment scheduling) and sensitivity of subject matter (e.g., HIV, mental health) appropriate for e-mail.

- Instruct patients to put the category of transaction in the subject line of message for filtering (e.g., PRESCRIPTION, APPOINTMENT, MEDICAL ADVICE, BILLING QUESTION).

- Request that patients put their name and patient identification number or date of birth in the body of the message.

- Configure an automatic reply to acknowledge the receipt of messages.

- Print all messages with replies and confirmation of receipt and place in patient's paper chart (or save copy of message to electronic record).

- Send a new message to inform patient of completion of request.

- Request that patients use the automatic reply feature to acknowledge reading provider's message.

- Maintain a mailing list of patients, but do not send group mailings where the list of recipients is visible to each patient. (Use blind copy feature in software.)

- Avoid anger, sarcasm, harsh criticism, and libelous references to third parties in messages.

- Ask patients to be concise and to limit one subject to each e-mail message.

Adapted from Kane B, Sands DZ. Guidelines for the clinical use of electronic mail with patients: the AMIA Internet Working Group, Task Force on Guidelines for the Use of Clinic-Patient Electronic Mail. JAMA. 1998;5:104-11.

handled by the practice. See Figure 18-1 for an example of a practice's e-mail policy.

Do I Have to Use Encryption Technology?

In 1997 the Committee on Maintaining Privacy and Security in Health Care Applications of the National Research Council released a comprehensive

Table 18-3 Medicolegal and Administrative Guidelines

- Consider obtaining patient's informed consent for use of e-mail. Written forms should:
  - —Itemize terms in communications guidelines.
  - —Provide instructions for when and how to escalate to a phone call or office visits.
  - —Describe security mechanisms.
  - —Indemnify the health care institution for information loss caused by technical failures.
  - —Waive encryption requirement, if any, at patient's insistence. (If patient is unable to use encryption and specifically requests from the provider not to use encryption, the request should be noted in writing by the patient and placed in the medical record.)
- Use password-protected screen savers for all desktop workstations in the office, hospital, and at home.
- Never forward patient-identifiable information to a third party without the patient's expressed permission.
- Never use a patient's e-mail address in a marketing scheme.
- Do not share professional e-mail accounts with family members.
- Use encryption for all messages when encryption technology becomes widely available, user-friendly, and practical.
- Do not use unencrypted wireless communications with patient-identifiable information.
- Double check all "To:" fields before sending messages.
- Back up e-mail into long-term storage at regular intervals (e.g., weekly); define "long-term" as the term applies to paper records.
- Commit policy decisions to writing and electronic form.

Adapted from Kane B, Sands DZ. Guidelines for the clinical use of electronic mail with patients: the AMIA Internet Working Group, Task Force on guidelines for the use of clinic-patient electronic mail. JAMA. 1998;5:104-11.

report on electronic confidentiality and security of health information (10). Soon thereafter, the Health Care Finance and Administration (HCFA), published its Internet security policy in November 1998 (11). HCFA's policy in regard to electronic communication has two requirements: a) an acceptable method of encryption (equivalent to Triple 56 bit DES) must be used to transmit sensitive HCFA data, and b) authentication or identification procedures must be employed by all users.

ELECTRONIC MAIL SERVICES
smg@med.stanford.edu

- Using Electronic Mail to Reach SMG
- Electronic Mail and Confidentiality

Using Electronic Mail to Reach SMG

Stanford Medical Group offers electronic mail access for several patient services. We would like you to consider using this method of communication when an immediate answer is not required. Examples include the following:

- Prescription renewals
- Non-urgent medical advice
- Routine HMO authorization requests
- Routine appointment scheduling
- Other non-urgent communication

Send electronic mail to the SMG mailbox: smg@med.stanford.edu

In subject area, list reason for message.
(prescription renewal, HMO authorization, routine appointment, medical advice, etc.)

In message body, please include your:

- Full name
- Medical record number
- Physician's name
- Return phone number

Message Processing

The electronic message will be reviewed and directed to the appropriate person to complete your request. You will receive an automatic reply letting you know the message reached Stanford Medical Group and then a confirmation by phone or electronic mail when the task has been completed.

E-mail, Patient Confidentiality, and Your Doctor

Stanford Medical Group has used electronic mail on a limited basis to communicate between physician and patients since 1994. We treat all communication between patient and physician with equal confidentiality whether by telephone, regular mail, or electronic mail.

Patient issues which we do not discuss via electronic mail are as follows:

- Protected diagnosis such as psychiatric conditions
- Results of HIV testing
- Work-related injuries and disability

(Continued)

Figure 18-1   Sample e-mail policy. (Courtesy Dr. Paul M. Ford, Stanford Medical Group.)

Electronic mail accounts maintained by commercial subscription services are usually considered private. When communicating from work, however, you should be aware that some companies consider electronic mail corporate property and your electronic mail messages may be monitored. *We recommend that you check with your company electronic postmaster before using electronic mail to communicate medical information or concerns with SMG.*

Project Goals

- Improve patient access to their physician and to SMG
- Facilitate coordination of medical care by generalist physicians
- Streamline administrative tasks

Comments and suggestions concerning the electronic mail service can be addressed to:

  Doctor's Name
  Doctor's E-Mail Address
  Doctor's Street Address
  Doctor's Telephone Number

Figure 18-1 (cont'd)

The confidentiality of medical information is of paramount importance to physicians and patients. However, current health information privacy standards are in an unsettled state. E-mail containing confidential information may be directed or intercepted by other online users such as employers. In its current form, unencrypted e-mail does not offer a secure means of communication. As the technology continues to evolve, user-friendly programs promise to secure electronic data on a global basis. In the meantime, a waiver of encryption requirement for patients should be used if patients do not wish or are unable to comply with the extra processing required for encryption.

Must I Have a Computer to Use E-mail?

Since the first electronic message was sent in 1969, e-mail has been the domain of the computer. In the last 2 to 3 years, however, other devices capable of e-mail have been introduced into the market. These devices, called info-appliances, include pagers, cellular telephones, e-mail machines, hand-held computers, and dedicated telephone systems. They are usually less expensive than a computer.

THE INTERNET

How Did the Internet Develop?

In the mid 1960s, the Advanced Research Projects Agency of the U.S. Department of Defense (ARPA), sponsored research into areas crucial to the successful linking of geographically remote computers to allow the sharing of data and resources. The first message sent over the ARPA-Net, the forerunner of the Internet, consisted of two letters, "L-O"; the network crashed as researchers were trying to type-in the word "log-on." In 1990, physicist Tim Berners-Lee developed software to link thousands of researchers to the particle physics laboratory CERN, in Geneva. His computer at CERN became the first Web content "server." He called his creation the World Wide Web (WWW).

How Many Physicians Use the Internet and for What?

A recent survey revealed that the number of physicians using the Internet almost doubled from 20% in 1997 to 37% in 1999 (12). The primary use of the Internet among physicians in 1999 was mainly for sending and receiving e-mail (Table 18-4).

What Are Patients Doing on the World Wide Web?

The Internet has witnessed a dramatic growth in the number of users. Currently, more than 65 million American adults (47% women) use the

Table 18-4　What Physicians Do When They Go Online

Online Activity	Percent
Send and receive e-mail	91
Access medical information sources	84
Collect travel information	80
Obtain product information	76
Communicate with professional associations	66

Source: 2000 AMA Study on Physicians' Use of the World Wide Web.

Table 18-5   Health-Related Information Retrieved by Consumers

Health-Related Content	Percent
Disease information (especially cancer and heart disease)	52
Diet and nutrition	36
Pharmaceuticals	33
Health newsletters	32
Women's health	31
Fitness	29
Children's health	15
Illness support groups	13

From Miller T, Reents S. The health care industry in transition: the online mandate to change. Cyber Dialogue Inc., 1998; with permission.

Internet. Where an adult goes on the Internet appears to be a function of gender. Men prefer to go to technology, sports, science, nature, and investing sites, whereas women show a preference for travel and recreation, health and medicine, food and cooking, and parenting and children sites (13). It is estimated that in 1999, 23 million adults searched online for health and medical information. Table 18-5 shows how consumers use the Internet for health-related content.

How Accurate Is Medical Information on the Internet?

With about 15,000 health-related sites on the Internet, health care professionals and patients alike have free access to an expanding volume of information that was previously difficult or impossible to access. Looking for useful and valid information on the Internet can be challenging because of the lack of control and speed with which the information is compiled.

The search engines (software programs that search the Internet for content related to a query) that scan the Internet for answers to questions use different technologies and respond differently to queries. The medicine-specific search engines are not necessarily the most reliable; in a study comparing the accuracy of different search engines (commercial engine, general engines, medicine-specific engine) in addressing 10 primary care clinical questions, the medicine-specific engine fared worst (14).

Table 18-6   Basic Standards for Judging Reliability of Information on the Internet

Basic Standards	Information Provided by Site
Content	Should describe clearly the scope, depth, accuracy, and reliability of information provided
Design, aesthetics, and ease of use	Web page layout (including graphics and the manner in which information is presented) should be easy to follow
Authorship	Should provide names of authors and contributors, credentials, and affiliations
Attribution	Should list all relevant references as well as sources of content
Currency	Should list dates when content was posted, frequency of updates, and site maintenance information
Disclosure	Should clearly display ownership, sponsors and developers of the Web site and identify sources of support and potential conflicts of interest

Physicians and patients should also be wary of the accuracy of information found on the Internet. A study (15) addressing the reliability of information on the WWW for parents seeking advice on managing a child's fever at home by comparing Internet advice with published practice guidelines (16) found that only 10% of the Internet sites provided reliable information.

These studies raise concern about the Internet's promise to provide accurate and reliable health-related information. Many instruments exist or are being developed that attempt to evaluate health information as it exists online (17). Until a uniform standard is developed, however, both health care professionals and consumers should use the standards shown in Table 18-6 (18,19).

How Might the Internet Affect My Practice?

Ultimately, the Internet will serve as a platform to link the key players in health care: patients, physicians, hospitals, health plans, payers, and service providers (Figure 18-2). The promise of integration and seamless operation is still, however, in its infancy. Portals such as www.zirmed.com for electronic claims services, www.speechmachines.com for electronic transcribing,

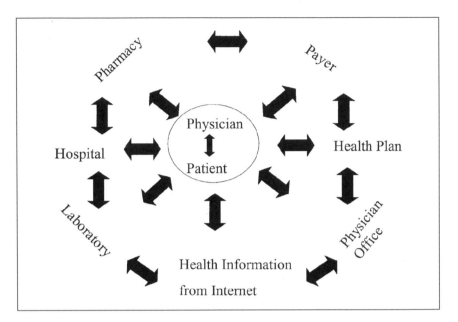

Figure 18-2 Futuristic model of connectivity and health care. (Courtesy Stanford Medical Group.)

www.personalmd.com for the patient medical record, and the recent purchase by Sun Microsystems of Web-based office suite software indicate that eventual integration is on the horizon. In May 1999, WebMD, a content and services site for consumers and physicians, announced plans of a merger with Healtheon, a connectivity site. One goal of this merger was to link electronically various health care participants onto a single Internet platform to streamline transactions and critical data transmission. One author has compared the medical role of the Internet to the rainforest canopy: from the ground one sees a collection of separate trees (the different systems, such as word processors, tracings, operative reports, laboratory reports), whereas from the canopy (the Internet user's view) one sees only a seamless web that allows creatures to move efficiently to reap its fruits without paying any attention to the individual trees (20).

Who Has Access to Confidential Patient Information?

Insufficient protection of patient-identifiable data may lead to discrimination and embarrassment for the individual. The delay in the integration of

**Table 18-7   Key Principles of Proposed DHHS Regulations
to Protect Health Information**

Boundaries	Health care information should be disclosed for health purposes only, with limited exceptions.
Security	Health information should not be distributed without patient authorization, and those who receive such information must safeguard it.
Consumer control	Persons are entitled to access and amend their health records and to be informed of the purposes for which they are used or disclosed.
Accountability	Those who improperly handle health information should be criminally punished and subject to civil recourse.
Public responsibility	Individual privacy interests must not override national priorities of public health, medical research, preventing health care fraud, and law enforcement in general.

Adapted from Hodge J Jr, Gostin L, Jacobson P. Legal issues concerning electronic health information: privacy, quality, and liability. JAMA. 1999;282:1466-71.

computerized health care data has been mainly caused by the challenge of protecting patient privacy rights. Current laws and regulations in the United States provide a fragmented set of federal and state laws that protect some types of individual health information but not others (21). After having failed to enact comprehensive health information legislation pursuant to the Health Insurance Portability and Accountability Act (HIPAA) of 1996, Congress has transferred that duty to the Department of Health and Human Services (DHHS), which has developed regulations that were first implemented in early 2000 (Table 18-7).

What Is the Internet-2?

A faster and more reliable Internet, the Internet 2, is already being tested. The Internet-2 is a consortium of more than 160 academic centers across the United States that are collaborating on making applications such as telemedicine, digital libraries, and virtual laboratories Internet-accessible. Another initiative, the "Next Generation Internet," is a government-sponsored program focused on developing applications and networking capabilities needed by federal agencies.

TELEMEDICINE

What Is Telemedicine?

Telemedicine refers to a wide range of telecommunication and electronic technologies that provide and support health care when distance separates the participants, such as telephone, radio, modem, facsimile, and video. In this chapter, *telemedicine* refers to communication between a host and a remote site using technologies other than telephone or fax.

Telemedicine is usually used for purposes of education or clinical consultation. It may be conducted in real time, such as interactive videoconferencing, or asynchronously, as e-mail or video is used. In addition to using monitors and cameras for receiving and transmitting video, digital diagnostic instruments (such as stethoscopes, sphygmomanometers, electrocardiographs, Doppler ultrasound, dermatoscopes, and pulmonary function devices) can be employed for consultation and education (22,23).

What Are the Clinical Uses of Telemedicine?

In addition to its most basic educational and clinical uses, telemedicine has been advanced through technological breakthroughs that have allowed the adoption of sophisticated digital instruments that assist in clinical examination and monitoring of patients. Telemedicine has already proven its feasibility in several challenging environments, including peacekeeping missions, the space shuttle, and remote areas such as the Arctic and Antarctic circles (Table 18-8). The fire department in San Antonio, Texas has installed interactive audio-video communication systems in some ambulances to facilitate interactions between the paramedics and emergency room physicians (24). The Veterans Health Administration has used telemedicine to support its operations in multiple ways, including teleradiology, telepathology, telecardiology, telemental health, teleradiology, teledermatology, telecare in diabetes, telenuclear medicine, tele-education, and for administrative purposes (25).

What Are the Limitations of Telemedicine?

Although the costs of the hardware and software are decreasing, establishing telemedicine is still quite expensive. Real-time video has not been

Table 18-8 Select Clinical Uses of Telemedicine

Correctional facilities	Providing health care to prisoners
Home care	Assisting care of patients through telephonically transmitted data and images
Emergency medicine	Facilitating EMS personnel interaction with physicians
Army medicine	Assisting medical decisions in wartime and peace time missions
Space missions	Assisting medical decisions on the space shuttle
Geographically remote areas	Bringing health care to geographically remote areas (e.g., Arctic and Antarctic circles)
Tertiary center consultations	Providing support from "super" consultants
Education	Fulfilling CME requirements and learning new methods in medicine and surgery
Robotics	Developing robots to aid with operating from a distance

perfected in most cases, although technological breakthroughs in bandwidth (information transmitted /unit of time) transmission have improved the quality of the picture. Other issues that still need to be ironed out before wide adoption of telemedicine include privacy and confidentiality of electronic medical information, medical liability and malpractice, reimbursement of clinical services provided, and licensure for interstate medicine (Table 18-9).

How Do I Structure My Practice to Include Telemedicine Capability?

The six-step approach formulated by Josey can be used to determine if telemedicine would improve one's practice (Table 18-10). First, the appropriate patient population must be identified because home telemedicine protocols are better suited for certain disorders (chronic airway obstruction, joint disorders, hypertension, heart failure, pneumonia, and cerebrovascular disease) (25). A significant number of these patients should have access to computers. The appropriate equipment should be selected and a business plan formulated, with expected expenses calculated. The

Table 18-9   Impediments to Wide Adoption of Telemedicine Technologies in Clinical Care

Start-up and support	Wide range of cost depends on purpose, use, and type of equipment purchased (range $5000–$100,000/site).
Licensure	Many states have limitations on out-of-state physicians who participate in telemedicine.
Reimbursement for services	Limited mostly to Medicare recipients in designated "health profession shortage areas."
Medical liability and malpractice	Current malpractice policies may not cover telemedicine contacts.
Confidentiality	Access to medical records must be carefully regulated to protect consumers.

service must be marketed to providers and patients, policies and protocols must be developed, and the office staff should be educated in the use of and indications for telemedicine. In addition, two principle criteria should be met before initiating telemedicine operations: 1) patients or caretakers must be able to operate the equipment reliably and safely, and 2) the patient should not require more than one intervention because the telemedicine may become less likely to succeed if it involves many steps (26).

INTERACTIVE TELEPHONY

Interactive telephony is a recent technology that allows communication between a computer and a touch-tone telephone. It is also known as *telematics* or *audiotex*. Users listen to recorded messages, confirm their selection from the menu by pressing the appropriate keys on their telephone keypad, and follow verbal instructions transmitted during the electronic conversation. For example, interactive telephones can deliver a prerecorded message about influenza vaccination to all patients in a practice who qualify for the vaccine. Patients may then select to hear further information by pressing certain numbers either to make an appointment or to learn more about the vaccine (2).

Interactive telephony is attractive to health care settings because of the ubiquitous presence of the telephone. Such systems can facilitate the

Table 18-10   A Six-Step Approach to Merging Telemedicine with Clinical Practice

Identify the appropriate patient population	Certain disorders are more suitable for home health telemedicine (e.g., chronic airway obstruction, joint disorders).
Select the appropriate equipment	Example: equipment needed to monitor diabetics (stop-motion video) is different from equipment needed to monitor patients with heart failure (digital stethoscope).
Formulate a business plan	Emphasize relevance of benefits to practice. Calculate expenses: hardware, software, support agreements, telephone line, staff training.
Market the service to providers and patients	Presentations and hands-on demonstrations capture audiences and provide a clear vision of telemedicine's capabilities.
Develop policies and protocols	For admission, discharge, patient safety, infection control, equipment malfunction, quality control, measures outcomes.
Educate staff	Focus on appropriate patient selection and equipment operation. Determine staff needs and plan educational content to meet their needs.

Adapted from Josey P, Gustke S. How to merge telemedicine with traditional clinical practice. Nursing Management. 1999;30:33-6.

function of a primary care physician in practice, both in terms of saving time and saving costs. One study (27) demonstrated that a computer telephony administered PRIME-MD instrument (a diagnostic instrument developed for identifying mental disorders by primary care physicians [28]) provided primary care physicians with an improved ability to detect mental disorders in their patients without additional physician time and at minimal expense. A small study of elderly patients showed that interactive telephony systems could improve medication compliance in this population (29).

REFERENCES

1. **Kuppersmith R, Holsinger C, Jenkins H.** The use of e-mail by otolaryngologists. Arch Otolaryngol Head Neck Surg. 1996;122:921-2.
2. **Balas A, Jaffrey F, Kuperman G, et al.** Electronic communication with patients: evaluation of distance medicine technology. JAMA. 1997;278:152-9.

3. **Jackler R.** Brave new world. Arch Otolaryngol Head Neck Surg. 1999;125:471-2.
4. **Carrns A.** Three words doctors dread: you've got mail. Wall Street Journal. 1999;Dec 9:B1, B13.
5. **Neill R, Mainous A, Clark J, Hagen M.** The utility of electronic mail as a medium for patient-physician communication. Arch Fam Med. 1994;3:268-71.
6. Doctors missing Internet health opportunity. 1999 CyberAtlas. Available at: ⟨http://cyberatlas.internet.com/big_picture/demographics/article/0,1323,5971_2 16411,00.html ⟩; accessed 5 December 1999.
7. **Havighurst C.** Unlocking the mailbox. American Medical News. 13 September 1999.
8. **Rettig R.** Health care in transition: technology assessment in the private sector. Santa Monica, CA: RAND; 1997.
9. **Kane B, Sands DZ.** Guidelines for the clinical use of electronic mail with patients: the AMIA Internet Working Group, Task Force on Guidelines for the Use of Clinic-Patient Electronic Mail. JAMIA. 1998;5:104-11.
10. **Committee on Maintaining Privacy and Security in Health Care Applications of the National Information Infrastructure.** For the record: protecting electronic health information. Computer Science and Telecommunications Board Commission on Physical Sciences, Mathematics, and Applications. Washington, DC: National Research Council; 1997.
11. **Office of Information Services.** Internet security policy. 1998, HCFA. Available at: ⟨http://www.hcfa.gov/security/isecplcy.htm ⟩; accessed 19 November 1999.
12. 2000 AMA Study on Physicians' Use of the World Wide Web.
13. **Miller T, Reents S.** The health care industry in transition: the online mandate to change. CyberDialogue Inc., 1998.
14. **Graber M, Bergus G, York C.** Using the World Wide Web to answer clinical questions: how efficient are different methods of information retrieval? J Fam Pract. 1999;49:520-4.
15. **Impicciatore P, Pandolfini C, Casella N, Bonati M.** Reliability of health information for the public on the World Wide Web: systematic survey of advice on managing fever in children at home. BMJ. 1997;314:1875-81.
16. **El-Radhi A, Carroll J.** Management of fever. In: Fever in Pediatric Practice. Oxford: Blackwell Scientific; 1994:229-31.
17. **Jadad A, Gagliardi A.** Rating health information on the Internet: navigating to knowledge or to Babel? JAMA. 1998;279:611-4.
18. **Silberg W, Lundberg G, Musacchio R.** Assessing, controlling, and assuring the quality of medical information on the Internet. JAMA. 1997;277:1244-5.
19. **Kim P, Eng T, Deering M, Maxfield A.** Published criteria for evaluating health related web sites: review. BMJ. 1999;318:647-9.
20. **McDonald C, Overhage M, Dexter P, et al.** Canopy computing: using the Web in clinical practice. JAMA. 1998;280:1325-9.
21. **Gostin L.** Health care information and the protection of personal privacy: ethical and legal considerations. Ann Intern Med. 1997;127:683-90.
22. **Perlin J, Collins D, Kaplowitz L.** Telemedicine. Hospital Physician. 1999;36:26-34.
23. **Grigsby J, Sanders J.** Telemedicine: where it is and where it's going. Ann Intern Med. 1998;129:123-7.
24. **Binius T.** Moving pictures: San Antonio leads the way in developing EMS telemedicine. American Medical News. 1999;March 22/29:28-9.

25. **Veterans Health Administration.** Telemedicine in Action. Available at: (http://www.va.gov/telemed/teleactn.htm); accessed 9 December 1999.
26. **Allen A, Doolittle G, Boysen C, et al.** An analysis of the suitability of home health visits for telemedicine. J Telemed Telecare. 1999;5:90-6.
27. **Kobak K, Taylor L, Dottl S, et al.** A computer-administered telephone interview to identify mental disorders. JAMA. 1997;278:905-10.
28. **Spitzer R, Williams J, Kroenke K.** Utility of a new procedure for diagnosing mental disorders in primary care: the PRIME-MD 1000 study. JAMA. 1994; 272:1749-56.
29. **Leirer V, Morrow D, Tanke E.** Elders' non-adherence: its assessment and medication reminding by voice mail. Gerontologist. 1991;31:514-20.

19

OFFICE MANAGEMENT ISSUES

Anna B. Reisman, MD

The main focus of the doctor on call has always been to resolve the problem and get back to sleep, whereas these nurses [after-hours triage nurses] are perfectly ready to talk for a half hour.

ROBERT YOUNG (1)

A patient's experience with the telephone colors his or her perspective of both the practice and the physician. The telephone may be the initial way the patient comes into contact with the practice and may be the mode in which the patient has the most contact with the practice. A positive experience on the telephone—polite and helpful receptionists, accessibility to medical personnel, quick return of calls—can mean a lot to a patient. Practices where receptionists are not helpful, where telephones are busy frequently or never answered, and where automated telephone trees are confusing and fail to cover the options most patients have in mind may be perceived as being inaccessible or uncaring and may ultimately lead to a loss of patients.

This chapter provides information on topics and challenges relevant to telephone medicine in the office setting (Table 19-1).

CHALLENGES

Many of the challenges in telephone medicine relate to the increasing volume of calls. Managed care, advances in medical knowledge, and the growing number of elderly patients have contributed to this increase. In managed care, requirements for referrals, medications, and preauthorizations generate a great number of calls. Managed care formularies often

Table 19-1	Topics Important in the Office Practice of Telephone Medicine
• Challenges of managed care	• Emergency department authorization
• Protocols	• Telephone fees
• Telephone policy information sheet	• Prescription refills
• Telephone hour	• Telephone triage
• Telephone type (cellular, cordless, other)	• After-hour services

change, leading patients to call with requests for new medications or non-formulary requests. Changes in managed care plans occur frequently, and patients may call requesting medical records or for refills of prescriptions from their former physicians.

Advances in medical knowledge have also increased the volume of calls; for example, more telephone calls are needed to monitor patients taking warfarin for atrial fibrillation. The aging of the population has led to more telephone calls between physicians and nursing homes, visiting nurses, and hospices. Direct marketing campaigns and media reports have led to more patients calling with questions about new medications and procedures.

TYPES OF TELEPHONE CALLS

Calls can generally be divided into symptom-related calls and non-symptom-related calls. The former includes patients calling with acute, subacute, or chronic symptoms. The latter includes administrative calls (e.g., calls regarding appointments, referral requests, insurance questions, prescription refills), and routine test results. As discussed in Chapter 1, administrative calls tend to be more frequent during the day and symptom-related calls are more common after-hours.

INSURANCE AND CALL DISPOSITION

Differences in reimbursement procedures in fee-for-service versus managed care and capitated settings can affect call disposition. In a fee-for-service setting, the physician may be more likely to suggest an in-person examination for a patient calling with a symptom, whereas in a capitated setting it may be more cost-effective for the physician to manage the problem over the telephone (see section below on Telephone Fees).

STAFF TRAINING

The office staff is on the front line of telephone medicine and can set the tone of the practice, yet most are trained "on the job" without clear guidelines regarding the disposition of telephone calls. Training workshops for office staff and nurses by practice management consultants are available and can be quite useful. It is also important for physicians to clarify their expectations to the office staff (2).

It is crucial that the staff be trained in how to assess when a call needs immediate physician attention. Without a clear standard, such messages from patients may end up in a number of different places: hand-delivered to the physician, on the physician's desk for immediate attention, or in an "in box" under a pile of other papers. In practices with nurses who triage calls, this assessment should be part of nurse telephone training. In practices where receptionists take messages without nurse screening, receptionists and other office personnel need to have some procedure for identifying important calls. An example of this is included in Chapter 4.

PATIENT INFORMATION SHEET

An instruction sheet on the office's telephone policy can be very useful for both doctors and patients. It should be part of the new patient package and should include information on procedures for emergency calls, routine calls during the day, calling for an appointment, calling for prescription refills, and after-hours calls (Table 19-2).

PRESCRIPTION REFILLS

A significant number of telephone calls are requests for prescription refills. To expedite such calls, which can be time-consuming, it is useful to

364 • *Telephone Medicine*

Table 19-2   Sample Telephone Policy Information Sheet

Emergency Calls
If you think the problem is life-threatening, call 911. If possible, call the emergency department (444-4444) before coming. The nurse or doctor there will be able to offer advice that may be important before you arrive.

If You Need a Same-Day Appointment
If you are sick and want to be evaluated that day, we recommend that you call first and make an appointment. You will be seen more promptly if you have an appointment. If you are unable to call first, you may come directly to the office, but you may have to wait to be seen.

Calling During Office Hours
Office hours are 8 a.m. to 5 p.m. weekdays. You will speak with a receptionist first who will help you if you need a medication refill or have a question about appointments. If you are calling with a symptom, a nurse will speak with you. If he/she is unable to help you, your physician will call you back as soon as possible. Calls received before 4:30 p.m. will be answered by the end of the same day. Calls received after 4:30 p.m. may not be returned until the next business day.

Calling When the Office is Closed
Calls that cannot wait until the next morning can be made to the physician on call (444-4445) from 5 p.m. until 8 a.m. every weekday, and 24 hours a day on weekends. An answering service will take the message and contact the physician on call, who will call you back within 15 minutes (sooner in case of an emergency).

Prescription Refills
Calls for prescription renewals should be made during office hours. Please call a week or two before your prescriptions run out, because we may not be able to provide prescriptions over the weekend or after-hours.

---

have a policy on prescription refills. In some office settings, a nurse or secretary records the refill request and it is given to the physician with the patient's chart. The situation is more complex in after-hours situations, since the patient's chart is usually not available. If the patient is not known personally to the physician, there are no records available, and the request is for a chronic medication, it may be appropriate to prescribe a limited amount until the patient can contact his or her regular physician in the morning or after the weekend. In some situations, it may be prudent to call the patient's pharmacy to check on the date and dose of the last filled prescription. Such a policy should be described to patients to discourage callers from calling during off-hours for prescriptions. An approach to patient requests for narcotics is discussed in Chapter 17;

Chapter 3 discusses medicolegal issues involved in prescribing medications over the telephone.

"TELEPHONE HOUR"

Some physicians incorporate a telephone hour into their daily schedules and request that patients make any necessary calls at that time. This has the potential to lessen the number of calls during the rest of the day. If the telephone hour is scheduled early in the day, it can be useful for the physician and office staff in planning the day's schedule by giving a sense of how many urgent visits will be needed (3). It can also increase patient satisfaction when patients know there is a time when they can reach the physician (3). However, it is also possible that not all patients who wish to speak with the physician will be able to get through.

PRIVACY AND CELLULAR PHONES

Cellular phones have made the physician-on-call's life easier by providing ready access to a telephone. In the age of caller ID, cell phones represent a means of keeping one's home number private. However, cellular phones (and home cordless phones) are easier to intercept, which may compromise patient confidentiality. A patient should be informed that the physician is speaking on a phone that may not be completely confidential. Alternatively, the physician may request that the telephone company block his home telephone number from caller ID systems. See Chapter 3 for a discussion on medicolegal issues of telephone confidentiality.

EMERGENCY DEPARTMENT AUTHORIZATION

Physicians whose patients have managed care insurance are sometimes responsible for granting authorization for emergency department (ED) visits as well as for hospital admissions. In teaching hospitals, the physician is often not expected to go to the ED when a patient is admitted and can handle most situations over the telephone. Most managed care plans put pressure on both patients and physicians to avoid using the ED as much as possible, because of the increased cost. However, the guidelines for managed care authorization for ED visits have been called "variable, vague, and unpublished" (4).

Because of the lack of reliable guidelines, the physician's role in these situations—at least in teaching hospitals, where there is housestaff available—is to evaluate the patient as completely as possible over the telephone and to make an informed decision. The physician can gather information about the patient from the evaluating physician or nurse in the ED and should speak with the patient if there is any doubt about the presentation or if reimbursement for the visit will likely be denied. It is interesting to note that one study found that almost half of patients denied authorization would return to the ED in the future, despite previous denials (5).

NURSE TELEPHONE TRIAGE SYSTEMS

History

Nurse telephone triage systems have been in existence since the mid 1980s, though health maintenance organizations (HMOs) began to use nurses to screen calls in the 1970s (6). Nurse triage arose from the need of physicians to delegate telephone calls, which were taking large amounts of time away from in-person patient care, to decrease costs and to eliminate unneeded office visits (6). The term *telephone triage* first appeared in Medline in 1990 and has since become a specialty of its own (6). Table 19-3 compares the advantages of nurse and non-nurse telephone triage systems.

Protocols

Telephone triage by nurses is generally based on a series of protocols that provide a standard approach to symptoms. The protocols are designed to identify conditions likely to require urgent or emergent care and to

---

**Table 19-3   Comparison of Nurse and Non-Nurse Telephone
Triage Systems**

Advantages of Nurse Telephone Triage
  Cost-saving in some settings
  Time-saving for physician
  Nurses generally better trained at telephone triage
  Nurses on shift: easier to return calls to patients, not tired in the middle of the night
  Decreased medicolegal problems from using good protocols and good
    documentation
  Increased patient satisfaction in some settings

Advantages of Non-Nurse Triage Systems
  Cost-saving in some settings
  Patients can speak with their own physician
  Patients have continuous access to their physician
  Physician knows patient better
  Increased patient satisfaction in some settings

---

provide self-care information when appropriate. Some protocols exist in
published form; others are computerized; some are HMO- or group-spe-
cific. Protocol systems usually have some form of physician back-up.

Many practices create their own protocols for use during office hours
in order to have more flexibility with patients who are already known.
Published protocols tend to be more conservative and more appropriate
for after-hours triage, when patients are less frequently known to the
triage nurse. Medicolegal issues related to nursing protocols are discussed
further in Chapter 3.

After-Hours Telephone Triage Services

Private After-Hours Services

Private after-hours services, or "call centers," provide trained nurses to
handle after-hours calls. These usually involve nurses who work from
one center rather than on-site (7). Managed care companies and private
practices who do not want to use their own nurses for after-hours calls
may choose to hire such services. A record of the call and the action
taken is usually faxed to the regular physician by the next morning. Some

services make and save a tape of every call in order to have records for possible lawsuits.

Recently, the American Academy of Pediatrics created telephone guidelines for such call centers in the pediatric setting, which may be generalizable to the adult setting. The guidelines included statements such as

- All patients of the subscribing practice should have access to the center.
- The staff of such centers should be well trained in telephone triage.
- A physician representing a subscribing practice should be on call at all times when the call center is being used as a backup for the following situations:
  a) when a caller insists on speaking with a physician
  b) when the nurse feels unable to handle a call
  c) when the physician wishes to be contacted for certain acuity levels
  d) when the nurse feels the caller will not comply with the advice (8)

Advantages of such services are that they relieve a practice of having to hire nurses for the night telephone shift, patients have access to telephone advice for 24 hours a day, and they can be cost-saving if the number of calls is high enough. Disadvantages include frequent busy signals and difficulty getting through to speak with a nurse.

Prerecorded Topics

Some after-hours telephone services provide a choice to the caller of either speaking with a nurse or selecting a prerecorded topic that matches the reason for calling. The Parent Advice Line, for example, offers recordings on 300 pediatric topics. TelMed in Connecticut is a service through the Yellow Pages that provides free recordings on medical and nonmedical topics. The availability of such information can decrease the number of calls to a live person by 30% (9). With the plethora of health-related information available over the Internet, however, prerecorded telephone topics may become obsolete.

Service Companies

Another variant of telephone triage service are service companies or subscription services, which provide 24-hour nurse coverage for telephone calls on a per-use basis. This can be cost-saving for smaller practices that may not want to invest in hiring nurses and equipment.

HMO Call Services

HMO-run call services are usually available 24 hours a day and tend to be highly computerized. Nurses have instant access to computerized patient records, including previous medical visits, recurring conditions, and medications with possible side effects, as well as computerized protocols and data on benefit plans and provider networks available to the patient. Sometimes the computer will prompt the nurse to the number of minutes elapsed before management of the problem, which many nurses do not like (6).

1-900 Numbers

In 1990, for profit telephone health care information service numbers received more than one million calls a month (10). They are used primarily by people without regular physicians—often the underserved and indigent. Such patients tend to go to the ED for serious problems but prefer not to go for less urgent symptoms. The 1-900 services will provide information but not a diagnosis. The Internet versions of 1-900 numbers are multiple nurse or physician consultation services on line. Some of these services are provided at no cost; others may be as much as $60 per hour or more.

NON-NURSE TRIAGE SYSTEMS

Group or private practices may employ a variety of non-nurse triage telephone call systems. Calls may come to the physician through an answering service, which provides the patient's name, number, and chief complaint. Some systems are often based on shared telephone call duty between several practice groups; in smaller practices, physicians may be on permanent call for their patients.

CHOOSING AN APPROPRIATE AFTER-HOURS SYSTEM

Deciding whether to use a nurse triage program, and which type, can be difficult. The physician must be comfortable with the idea of nurses handling patient calls. A physician in private practice whose patients have been able to reach him day and night in the past may find some resistance to changing over to a nurse telephone triage system. Using one's own nurses to handle calls during the day and after-hours can be easier, because the nurses are known to the physician and the physician can speak with patients if necessary. Using a private service with nurses of unknown quality can be more risky.

Confidentiality is another important issue. Nurses whom the physician does not know personally must be trusted with confidential patient information. Health insurance companies that pay for telenursing programs may have access to patient information.

Once the decision is made to use nurse triage, one must choose between setting up one's own after-hours telephone system and hiring an existing after-hours service. Patient volume, flow, number of calls, and the types of calls that require the most time should all figure into the decision. The telephone company can be helpful in determining how many lines will be needed based on the expected number of incoming calls. It is also useful to track which types of calls are the most common in the practice, as this can help structure the training for staff (11). A patient survey on telephone needs or feedback on an existing system can be helpful both before and after initiating a telephone system.

CHARGING FOR TELEPHONE CALLS

Pros and Cons

Should physicians receive compensation for telephone calls? This question has been debated for many years. The following pros and cons relate to a fee-for-service environment. Proponents of telephone fees say that the risk assumed by physicians—just as for other professionals—deserves to be compensated. They stress that physicians would likely feel less exploited by callers, with a decreased likelihood of frequent calls for trivial or minor complaints (12). Patients might value calls more, seeing them as "not a favor but a professional treatment" (13). Physicians would likely document calls better and would have an incentive to follow up on calls to monitor active problems (12). More office time for calls would likely be protected for calls, which might save money as well (13).

Those against fees for telephone calls to physicians believe that the cost to both patients and physicians would be high. For patients, there would be increased medical costs; for physicians, the cost, perhaps, of their patients' trust. The common patient perception that physicians are motivated by greed may be enhanced (14). Though telephone fees might discourage calls for minor complaints, they may also cut back on appropriate and important calls, because patients often call because they are unsure of the import of their symptoms (12,14). Similarly, calls that seem minor to the physician may be of great importance to the patient (14). It would be difficult to create a payment scheme for the many calls that are made for reassurance rather than for an acute medical problem (13). Finally, there would be room for abuse in such a system; some physicians might make frequent, unnecessary calls to boost their income.

For managed care and other capitated plans, increased telephone use has the potential to save costs. Payment for telephone service is included, or "bundled," with the reimbursement for the office visit, so physicians in managed care generally have a disincentive to encourage frequent in-person visits.

Physician Organization Statements

Several physician organizations have made statements in favor of telephone compensation for physicians. In 1990, the American Medical Association's Council on Medical Service report stated that physicians should be able to be reimbursed for telephone calls to patients. The same year, the American Society of Internal Medicine made a policy statement stating that "telephone services that are reasonable, properly documented, and of high quality should be billable services that merit reimbursement by patients and third parties" (15). More specifically, the problem evaluated on the telephone must involve a new diagnosis or new treatment; the telephone evaluation must save the patient from an office or ED visit; and the physician must handle the call personally. Examples would include treating relapses of previous conditions when able to over the telephone, if significant time and judgment are needed; reporting laboratory results that require a significant change in medication or other tests; or extended family counseling by telephone.

Nonreimbursable calls would include calls that are generally no more than 2 minutes long and can often be delegated to staff after brief instructions, with minimal documentation, liability potential, and patient counseling (15). Some examples of nonreimbursable calls would include reporting normal test results, a call from a patient who is seen immediately after as a result of the call, a brief discussion to confirm stability of a chronic problem and continuity of present management, and confirming resolution of an acute problem without any complications.

HCFA Policy

In 1991 the Health Care Financing Administration (HCFA) regulations provided three current procedural terminology (CPT) codes for telephone calls:

- **99371**—for simple or brief calls (laboratory results or clarification or alteration of previous instructions or therapy)
- **99372**—for intermediate calls (advice to an established patient on a new problem, initiation of therapy, detailed discussions of test results or future care plans)
- **99373**—for complex or lengthy calls (counseling anxious patients or prolonged talks with patients or family members about complex services necessary for a serious illness) (16)

Despite the existence of these codes, most Medicare carriers will not pay for telephone services (16). Medicare policy states that telephone calls before and after an office visit are part of a physician's work during that office visit. For patients with managed care insurance, services have a set fee, and unbundling is a policy violation (16).

Medicare will allow for separate reimbursement when the physician must obtain medical information from a previous physician or hospital by telephone; however, this will only be reimbursed if it is accepted practice

ONE PHYSICIAN'S TELEPHONE FEE SYSTEM

One internist in private practice initiated a system of telephone fees in her practice setting. She created two types of fees. *Disincentive fees* were charged for "unwanted" calls, such as calls from patients who telephoned before their first visit, doctor-shoppers, excessively long calls, callers who insisted on speaking with the physician for straightforward issues, callers who misrepresented the urgency of a call, and callers who refused to divulge the reason for the call to the office staff. She would decide if a call met these criteria and inform the patient. She believed that charging for such calls was preferable to the "frank gruffness" that many physicians use when they feel a call is not appropriate (13).

Intensity-of-service fees, which were higher than the disincentive fees, were used when a major change in the patient's management was made over the telephone. Such changes included managing multiple problems, dictating a letter, or conducting intensive pre- or post-case review.

Implementing such a system, the physician has written, involves first overcoming one's misgivings about charging. She suggested clarifying the telephone services and restrictions to patients, informing them of the potential fee, and letting them know that any call of an emergency nature that could be dealt with over the telephone would be taken gratis, that the fee would be waived for brief calls that the physician wants to encourage (refills, lab results, messages quoted by nurse or secretary), and that in the case of financial hardship the fee would be reduced or waived.

in the community to do so (15). In addition, Medicare reimburses for telephone management of anticoagulation therapy when changes are made in warfarin dosing (15).

REFERENCES

1. **Weinstein S.** On call again? With a triage service, relief is in sight. Physician's Practice Digest. 1998:3.
2. **Curtis P, Evens S.** The telephone interview. In: Lipkin M, Putnam S, Lazare A, eds. The Medical Interview: Clinical Care, Education, and Research. New York: Springer-Verlag; 1995:187-95.

3. **Curtis P.** Improving the use of the telephone. The Practitioner. 1991;235:514-20.

4. **Zautcke J, Fraker L, Hart R, Stevens J.** Denial of emergency department authorization of potentially high-risk patients by managed care. J Emerg Med. 1997;15:605-9.

5. **Chan T, Hayden S, Schwartz B, et al.** Patients' satisfaction when denied authorization for emergency department care by their managed care plan. J Emerg Med. 1997;15:611-6.

6. **Wheeler S.** Telephone Triage: Theory, Practice, and Protocol Development. Albany, NY: Delmar Publishers; 1993.

7. **Poole SS, Schmitt BD, Carruth, T, et al.** After-hours telephone coverage: the application of an area-wide telephone triage and advice system for pediatric patients. Pediatrics. 1993;92:670-9.

8. **Rose V.** AAP report discusses success factors for pediatric call centers. Am Family Physician. 1999;60:1242-5.

9. **Schmitt B.** Pediatric Telephone Advice, 2nd ed. Philadelphia: Lippincott-Raven; 1999.

10. Telephone health information services are popular. Hospitals. 1991;65:14.

11. **Katz H.** Telephone Medicine Triage and Training: A Handbook for Primary Health Care Professionals. Philadelphia: FA Davis; 1990.

12. **Sorum P.** Compensating physicians for telephone calls. JAMA. 1994;272:1949-50.

13. **Braithwaite S, Unferth N.** Phone fees: a justification of physician charges. J Clin Ethics. 1993;4:219-24.

14. **Isaacman D.** Telephone fees: are they worth it? J Clin Ethics. 1993;4:271-3.

15. ASIM Policy and Guidelines on Appropriate Use of the Telephone for Diagnosis and Treatment of Patients. Washington, DC; 1990.

16. **Barton EB, Brown JL, Curtis P, Lichtenfeld L.** Making phone care good care. Patient Care. 1992;Dec 15:103-18.

20

· · · · · · · · · · · · · · · ·

TEACHING TELEPHONE
MEDICINE

· · · · · · · · · · · · · · · ·

David L. Stevens, MD

CASE 20-1

It is July 5 at 10 p.m., and Dr. Adderly, fresh out of residency, is on call for the practice he's just joined. He receives a call from Mrs. Vaughan, a 68-year-old patient of one of his colleagues.

Mrs. Vaughan: "Doctor, I'm having terrible diarrhea. It won't stop, and I can't take it anymore. What can I do?"

Dr. Adderly immediately pictures a frail and volume-depleted woman on an emergency department stretcher receiving intravenous normal saline with a nurse taking vital signs every 30 minutes. His heart races. He wants to ask her what her vital signs are. Without those, how can he know what to do? His instinct is to refer her to the emergency department, where he can examine her, get laboratory tests done, and give her prompt treatment.

Dr. Adderly: "I think it would be best if you went to the emergency department right away. I can meet you there in 20 minutes."

CASE 20-2

On Friday evening at 5:30, as she is putting on her coat to leave, Dr. Holiday gets a call from one of her long-standing patients, Mr. Young.

Mr. Young: "Doctor, I've had this scratchy throat all week and my head's stuffed up. I've tried everything, but last time the only thing that helped was antibiotics. The drug store said you could just phone it in. I have the number for you."

Dr. Holiday, feeling her blood pressure rise, weighs her options: begin a potentially frustrating, time-consuming effort to explain that antibiotics should not be prescribed until a throat culture has been done and the results are shown to be positive, or just phone in the prescription.

After letting him know how exasperated she is that he waited until Friday to call, she makes a half-hearted stab at the first option, then, demoralized, wraps up the phone call with the second option.

IMPORTANCE OF IMPROVING SKILLS IN TELEPHONE MEDICINE

Before beginning practice, physicians have the benefit of years of supervised practice. Virtually all of this supervision, be it in medical school, residency, or fellowship, is with patients seen in person. One might guess that physicians don't need specific training in telephone medicine. Many physicians and patients would not identify the telephone consultation as a primary role of the physician. Telephone calls from patients are much shorter than in-person visits, and physicians seldom charge patients for this service. However, telephone consultations are in fact one of the primary roles of physicians, both in terms of the quantity of the workload and in the scope of medical care that is provided (see Chapter 1). Moreover, as the above cases demonstrate, there are critical differences in the fundamental approach on the telephone.

Perhaps the most persuasive argument for training in telephone management is that physicians do not consider themselves competent in this realm. A study of practicing internists found a low level of confidence in practicing telephone medicine and a desire for formal teaching in telephone medicine (1). Another study (2) found that in nearly one-third of calls, the physician perceived the main reason for the call differently from the patient (see Chapter 1).

Although most residency directors believe training in telephone medicine is important, only 6% of internal medicine residencies were teaching telephone medicine in 1995 (3). The good news is that more residencies are teaching telephone medicine, and there is some evidence that focused

Table 20-1   Outcomes Improved by Training in Communication Skills
• More complete history taking
• Greater appreciation of patient's need for help
• Fewer calls resulting in physician's negative emotional response

training can help physicians improve their performance and their attitudes (Table 20-1) (4-6).

FOCUS OF CHAPTER

Objectives

This chapter will first explore learning needs in telephone medicine and provide a list of objectives for teaching telephone medicine. There will be a discussion on general approaches to teaching telephone medicine, incorporating the medical education research by Skeff et al (7). This will include discussion of strategies for developing the learner's sense that this endeavor is important and relevant to his or her practice as well as experiential activities to maximize the development of relevant skills. Finally, a curriculum will be discussed in detail in which four complementary activities will help the learners develop their knowledge, skills, and attitudes to allow them to improve their practice of telephone medicine. These activities include: 1) discussion of personal experiences/challenges in practicing telephone medicine, 2) critique of a "trigger tape" telephone interview, 3) practice interviews/role plays focusing on specific challenges, and 4) discussion of documentation.

Size of Group

This curriculum includes a series of focused discussions and practice exercises to allow the participants to determine their own learning needs and to start addressing them. These types of interactive activities work best when the group is small enough to allow everyone to be involved (anywhere from two to 20 participants). Groups much larger than this may

experience diminishing levels of participation, although experienced facil-
itators may still be able to accomplish adequate participation.

Time Requirements

This curriculum is designed to fit into any time structure, but a minimum
of 1 hour per session is recommended. Depending on the goals of the participants, the group may opt for multiple sessions to allow more topics to
be covered.

Role of the Facilitator

This curriculum is designed to be self-sufficient: no telephone medicine
"expert" is required. However, it is helpful to designate a facilitator for
the various activities. The facilitator should have some experience with the
types of educational methods used here, especially role play and review of
a trigger tape. Recommended strategies for facilitating these activities are
included below.

WHAT ARE THE LEARNING NEEDS
IN TELEPHONE MANAGEMENT?

Much of this book has been devoted to providing a knowledge base that is
helpful in practicing telephone medicine. However, in addition to a solid
knowledge base, a core set of skills is essential for handling the wide range of
problems confronted on the telephone, as discussed in detail in Chapter 2.
Equally important is attitude. Without a positive set of attitudes, a skilled
and knowledgeable physician is less likely to do his or her best work.
Moreover, learners need to perceive that the material (i.e., telephone management) is relevant in order to be willing to challenge themselves to work
on improving their abilities (7).

As the cases at the beginning of this chapter suggest, the content of a
training session should be tailored to the specific needs of the participants.
Residents or attending physicians just beginning practice such as Dr.
Adderly may have little confidence in their abilities to take an adequate
history over the telephone. More experienced physicians such as Dr.

Holiday may be confident and competent in this realm but be uncertain how to manage patients with common colds who demand prescription for antibiotics. For a telephone medicine training session to be maximally effective and efficient, a set of specific learning objectives should be decided on ahead of time (see below, How Should Telephone Medicine Be Taught). The following discussion and Tables 20-2 through 20-6 provide a starting point to begin selecting specific skills and attitudes to work on.

Attitudinal Learning Objectives in Telephone Medicine Training

As Cases 20-1 and 20-2 demonstrate, constructive, positive attitudes are essential in telephone medicine practice. A physician such as Dr. Adderly who doesn't have confidence in his own ability to take a thorough history may needlessly refer low-risk patients to the emergency department. A physician such as Dr. Holiday may not trust her ability to gain the confidence of a demanding patient and to provide counseling on the lack of effectiveness of antibiotics for the common cold. She may just "give up" and call in the prescription without even giving the challenging task her best effort. Table 20-2 provides a partial list of some attitudes that are essential for physicians to do their best work over the telephone. Depending on the type of practice and level of experience of the physician, these objectives can be tailored to suit individual needs.

Table 20-2   Attitudinal Learning Objectives

The physician/student will:

- Demonstrate interest in improving abilities in practicing telephone medicine

- Demonstrate a commitment to conducting telephone consultations with the same level of thoroughness and careful judgment as he or she would in any other medical setting

- Show willingness to explore patients' hidden concerns

- Show consideration for callers' medical and psychological distress

- Express a commitment to document telephone encounters appropriately

- Express increased confidence in his or her ability to provide high-quality telephone medical care

---

Table 20-3   Data Gathering Learning Objectives

The physician/student will:

- Address barriers specific to telephone communication

- Allow patients to give a full description of the problem without interruption

- Encourage a complete, chronological description ("narrative thread") of the chief concerns

- Gather relevant information on past medical history, medications, allergies, etc.

- Gather pertinent social history in a sensitive manner, including illicit drug use and sexual history

- Inquire about "pertinent negatives" for ruling out emergency/urgent conditions

- Note cues of underlying/hidden concerns

- Survey patient's underlying/hidden concerns (the "actual reason for calling")

- Determine the patient's assessment of the situation

Table 20-4   Therapeutic Rapport Learning Objectives

The physician/student will:

- Explore the patient's underlying concerns and express understanding to the patient

- Handle strong emotions (anger, fear, etc.)

- Make empathetic statements in response to patients' statement of suffering/strong emotion

- Adopt a manner of speaking (tone, volume, rate) that conveys caring

---

Skills Essential to the Practice of Telephone Medicine

Cases 20-1 and 20-2 illustrate that communication skills should be a primary focus of telephone medicine training. The cases above illustrate the need for competence in data gathering, developing therapeutic rapport, and patient education/negotiation on the telephone. In addition, telephone management's increased focus on triage and patient self-care requires an approach to decision making that is distinct from that of in-person management.

Tables 20-3 through 20-6 provide a list of some of the learning objectives that should be considered by physicians seeking to improve their

---

Table 20-5   Patient Education/Negotiation Learning Objectives

The physician/student will:

- Discuss his or her assessment and management options with the patient in understandable language
- Deliver difficult news (e.g., abnormal lab results, need to go to emergency department immediately) clearly and sensitively
- Survey the patient's management preferences
- Repeatedly check the patient's understanding of the situation and what the doctor has said
- Manage unreasonable demands respectfully
- Negotiate a plan (including if, when, and where in-person evaluation will take place)
- Assess the feasibility of the plan
- Inquire about what the patient intends to do

Table 20-6   Decision-Making and Other Learning Objectives

The physician/student will:

- Assess the need for emergency/urgent in-person treatment
- Explore the full range of home management strategies available, including self-monitoring
- Document the essential features of the telephone consultation

---

competence in communication and decision-making skills essential to practicing telephone medicine. The activities accomplished with these skills is generic to all medical interviews, but the specific skills necessary on the phone can be quite different, as the cases above illustrate (see Chapter 2).

Knowledge Essential to the Practice of Telephone Medicine

The knowledge base required to practice telephone medicine effectively is broad. Moreover, physicians' knowledge requirements vary with the nature of their practices. There are, however, a number of general points that should be known by anyone practicing telephone medicine (Table 20-7). Awareness of these will help participants raise their expectations for the

---

Table 20-7   General Knowledge Learning Objectives

The physician/student will:

I. Know that telephone medicine is common and important to medical practice
   A. Describe the evidence that telephone medicine is a frequent part of practice
      1. 25% of medical encounters are on the phone (8)
      2. Telephone medicine may become more common: managed care programs often mandate telephone access (% may be increasing)
   B. List the diverse uses for the telephone in medicine practice

II. Describe the evidence that telephone medicine training is important (1)
   A. Practicing internists are not confident in practicing telephone medicine
   B. Practicing internists believe telephone medicine should be taught formally during residency

III. Know that practicing good telephone medicine is similar to practicing good medicine in other settings
   A. List the essential tasks/skills in telephone medicine practice
      1. Obtaining a good history (including eliciting hidden concerns)
      2. Clinical reasoning
      3. Exploring options for management
      4. Negotiating a plan with the patient
      5. Follow-up
   B. Physicians already require competency in different settings with different limitations (inpatient vs. emergency department vs. community-based practice)
   C. Identify the necessary elements of documenting a telephone consultation
   D. List the reasons for documentation of telephone encounters, including legal aspects and continuity of care

IV. Discuss distinct challenges in telephone medicine
   A. List common barriers and challenges in the telephone consultation
   B. Name some solutions to these barriers

---

quality of their telephone medicine practice and may help them focus on learning telephone medicine. Additionally, they may learn from each other what strategies are effective in improving patient care on the telephone. As with skills and attitudes, some of the knowledge points in Table 20-7 will already be known to more advanced learners, and the facilitator should focus on the needs of the group.

Additionally, there is specific knowledge that is important for the management of medical problems over the phone. In Case 20-1, for example, Dr. Adderly should be familiar with the "warning signs" in acute diarrhea that would help determine the need for emergency department referral (see Chapter 6). In Case 20-2, Dr. Holiday would benefit from the knowledge of what historical elements are useful in determining the presence of strep-tococcal pharyngitis, as well as which approaches are likely to be effective when negotiating with a demanding patient (see Chapters 10 and 17).

HOW SHOULD TELEPHONE MEDICINE BE TAUGHT?

Research from the fields of education and psychology provides evidence that learning will be maximized if the learning activities are group directed and experiential. Attitudinal learning will take place if the participants value the material and if the content is designed to meet specific needs that are essential to their medical practice (7). Skills development is likely to be more effective and time-efficient if the activities are participatory and include feedback. Knowledge is often most efficiently conveyed by assigned readings or a brief lecture/discussion. Discrete learning objectives should be defined to ensure that the teaching is effective (7).

The discussions below address the general approaches for meeting these learning objectives and introduce the key activities that are likely to meet the learning objectives. After this, the individual activities that comprise the curriculum are discussed in depth.

Approaches for Attitudinal Learning

As stated above, the first attitude necessary in learning to practice telephone medicine better is simply the willingness to improve one's abilities in practicing telephone medicine. Without this attitude, busy residents or attendings will resent their time being taken away from what they might see as more vital activities, be it learning other topics or just catching up on sleep. Enthusiasm for improving one's abilities in telephone medicine can be fostered by reviewing previous experiences, which will remind them of what they personally find challenging, and by group discussion, in which the group can provide positive reinforcement of each individual's learning needs (7). A third approach is problem-based learning. Adult learning theory suggests that learners confronted with a specific challenge, such as those in Cases 20-1 and 20-2, will feel the urge to improve their abilities more acutely than if the topic is discussed in more general terms (7).

The above approaches are also effective at achieving the other attitudinal learning objectives in Table 20-2. For example, discussing a participant's experience interviewing a caller in need of emergency contraception will promote a consideration for a caller's possible psychological distress, and discussing a caller with chest pain is likely to lead to a discussion of the importance of documentation.

Approaches for Learning Skills Development

The approach to developing skills in telephone communication and decision-making should include a chance to think about and practice some specific skills in the telephone interview with feedback on how these skills were performed. Modeling, in which the learner actively considers the skills necessary for caring for the patient (7) can be accomplished by listening to and discussing a trigger tape of a telephone interview, and practicing the telephone interview with role play and/or a standardized patient. Trigger tape review has the advantage of allowing the individual participants to benefit from the group's greater wisdom and allows the physician/student to think critically without being confrontational (you can't hurt the tape's feelings!). On the other hand, practice interviews allow the individuals to try new behaviors in a safe environment. For example, skill in detecting cues of hidden concerns can be developed by listening to a tape of a patient whose worry about her dysuria seems out of proportion to the degree of physical discomfort (see Appendix I, page 401). Thoroughness in gathering past medical history can be developed by role playing a case of a patient with HIV and diarrhea (especially if the patient doesn't tell the physician about the HIV until asked specifically about "other medical problems"). Practice interviews also allow for the provision of feedback on the participant's performance. The feedback can be directed specifically to the skill being focused on. This may allow the participant to correct ineffective behaviors immediately, and may promote increased confidence in the participant's abilities.

Approaches for Knowledge Base Development

As Table 20-7 demonstrates, the general knowledge required for learning and practicing telephone medicine is essentially limited to the facts that telephone medicine is important and telephone medicine requires train-ing. Many experienced physicians will already know this, but summar-izing the facts, especially the study results on the frequency of calls and physicians' low levels of confidence, may be helpful in getting the par-ticipants "on board." In these cases, the facts speak for themselves and can be presented simply either in a handout or in a brief lecture. Knowledge of how to address barriers may often come from other partici-pants, so discussions may also meet the knowledge-development objec-tives. Knowledge of specific approaches to telephone problems can be taught from sources such as this book and telephone-based studies of medical issues.

THE CURRICULUM

With these approaches in mind (reviewing previous experiences, group discussion, problem-based learning, and modeling/practice with feed-back), telephone medicine training should include a variety of participa-tory and group-directed activities. Table 20-8 summarizes these activities and the learning objectives they are designed to accomplish.

Preparation: Survey of Participants' Most Challenging Experiences/Facilitator Prepares Relevant Practice Cases

Objectives

Attitudinal—Participants will:

- Identify areas that they need to work on
- Demonstrate interest in improving telephone medicine abilities
- Identify learning needs/skills that are directly relevant to clinical needs and interests

Depending on the participants' level of experience, the facilitator may choose to take some simple steps in advance to achieve the group's "buy-in" to learning telephone medicine. A simple questionnaire (Table 20-9) or informal survey asking the participants in advance about what they find

Table 20-8   Summary of Each Activity and the Suggested Time Allotted

Activity	Time Allotted (total: 1 hour)

Preparation
  Survey of participants' most challenging experiences
  Facilitator prepares relevant practice cases
	Before session

Discussion of participants' experiences and challenges	15 min
Trigger case/discussion	15 min
Skills practice/role play	25 min
Documentation/discussion	5 min

Table 20-9   Sample Questions for Surveying the Participants' Needs

- *What one or two patient phone calls did you find clinically challenging or frustrating?*
- *What type of phone calls do you find the most difficult?*
- *Which aspects of telephone medicine would you like to learn more about?*

challenging will heighten the participants' sense of the relevance of the material. It will also allow the facilitator to select or design practice cases that are relevant to the group's needs (see below for discussion of case/role play design).

Part 1: Discussion of Participants' Experiences

Objectives

  Knowledge—Participants will:

- List the diverse uses for the telephone in medicine practice
- List common barriers and challenges in the telephone consultation
- Name some solutions for these barriers

  Attitudinal—Participants will:

- Identify additional areas for development through listening to other participants' challenges

- Identify areas that they need to work on
- Demonstrate interest in improving telephone medicine abilities
- Identify learning needs/skills that are directly relevant to their clinical needs and interests

A relaxed and enjoyable "ice-breaker" for a group is to discuss individual experiences with telephone medicine—both the positive and the negative. The process for this discussion is summarized in Table 20-10. Participants should be encouraged to define how the encounters they describe affected the patient's care, what barriers were confronted, and what the clinical challenges were. As this occurs, the group can begin to focus on the broad topic of telephone medicine and develop an enthusiasm for working on it.

Individual participants, hearing each other's stories, will pick up new ideas on how the benefits of telephone medicine can be used to improve their patient care (Table 20-11). Additionally, the group will begin to see common barriers and clinical challenges. The most pressing of these should be identified for use in the role play scenarios. Time permitting, the group can brainstorm solutions for common barriers, such as those in Table 20-12.

Table 20-10   Process for Discussion of Participants' Experiences and Perceived Benefits

1. Each participant (time permitting) briefly describes a telephone medical encounter, discussing:
   - Probable effect of the encounter on the patient's health
   - Any barriers to effective care
   - Any clinical challenges

2. Discussion of cases:
   - Facilitator or other participant keeps a list (on a blackboard, flip chart, or overhead) of all the benefits, communication barriers, and clinical challenges mentioned
   - Common themes should be highlighted after all participants have spoken; the list of potential benefits in the practice of telephone medicine can be quite long, so the group should try to think broadly (see Table 20-11)

3. Facilitator may select one or two common communication barriers from the list and ask the participants to brainstorm possible solutions (see Table 20-12)

Table 20-11   Benefits of Telephone Medicine

Improved Access to Care
- Allows patient access to physician for intercurrent problems: advice/treatment/disposition
- Gives patient help with decisions at home

Augmentation of Office Visit
- Follow-up on issues from visit: toleration treatment/efficacy, reminder (e.g., smoking cessation)

Doctor-Patient Rapport
- Increases patient satisfaction
- Heightens sense of partnership with patient
- Allows patient to communicate with physician from more comfortable environment

Reduction in Cost/Morbidity
- Early diagnosis of problems
- May help patient avoid emergency department visit
- May improve quality of care/outcomes

Financial Benefits
- May become reimbursable
- May improve efficiency in managed care/capitated setting

Part 2: Trigger Case Critique

Objectives

Knowledge—Participants will:

- List important behaviors in the telephone consultation
- Identify effective and ineffective practices in telephone consultation

Attitudinal—Participants will:

- Express a sense of the importance of effective communication in producing a good clinical outcome

Skills—Participants will:

- Recognize cues of hidden concerns

Table 20-12   Common Barriers in Practicing Telephone Medicine and Possible Solutions

Barrier	Solution
• Information gathering	
No physical exam	More thorough history
Cannot "eye-ball" patient	Ask family ("Does your husband seem very sick to you?")
No "non-verbal" cues	Pay attention to voice tone, speech rate, language
No medical chart/don't remember patient	Ask patient/family for more information
• Doctor's personal barriers	
Doctor too busy	Call patient back, have back-up doctor available
Doctor on vacation	Cross-coverage system
Documentation is inconvenient	Establish system for getting note into chart
• Medical knowledge/experience	
Lack of experience	Check with more experienced doctor and call back; use books or other sources
No books	Same as above
• Patient's personal barriers	
Language barriers	Use family to interpret; call back patient with interpreter; use telephone company interpreting service
Patient doesn't know physician/ no existing relationship	Use relationship-building skills

The next step is to begin to think critically about the telephone interview, considering which behaviors are effective and which are ineffective and to begin to list the discrete skills that are necessary for managing a patient over the telephone. An activating method to accomplish this is to listen to and critique a telephone encounter between a physician and a patient. Ideally, a tape of this interview should be made in advance (see Appendix I), but the script can also be read aloud. The goals and process for this exercise are summarized in Table 20-13.

Summary of the Case

Depending on the needs and level of the group, the facilitator can use the script in Appendix I or write a case. In this case, an inexperienced

---

Table 20-13   Trigger Case Critique: Recommended Process

1. Participants should be prepared to note the following as they listen to the encounter:
   - *What did the physician do in the interview that was effective in assessing and managing the patient?*
   - *What did he or she do that was not effective?*

2. Play the tape (or have three participants read the script in character)

3. Discuss: What did the physician do that was effective? *Remember to discuss his or her effective behaviors fully before listing the ineffective behaviors.* Consider having someone take notes on the group's observations.

4. Discuss: What did the physician do that was *not* effective? *Focus on behaviors, not interpretations* (e.g., *"His tone was very monotonous"* rather than *"He sounded like he didn't care"*).

---

physician receives a call from a woman with acute dysuria. The physician takes her history, makes a working diagnosis, and counsels the patient on the probable diagnosis and management options. The physician also addresses the possibility that the patient might be concerned about having something dangerous. Although the physician demonstrates a number of positive practices and skills, he also falls short in a number of ways. Some of these shortcomings are obvious, and some are more subtle.

Discussion of Effective Behaviors

It is often easier for participants to jump right to "shooting down" the physician with a list of his shortcomings, but the group can learn a lot from identifying all the physician's positive behaviors as well. The group should try to name a broader task addressed by each specific behavior. For example, a specific behavior was asking about the patient's last menstrual period. A broader task addressed by this behavior might be ruling out a condition that is more of an emergency (in this case, ectopic pregnancy). Listing broader tasks will allow more thorough critiquing because the group can then discuss whether that task was fully accomplished. For example, were all possible emergency conditions ruled out? Table 20-14 lists some of the effective behaviors of the physician in a sample trigger case (Appendix I, page 401).

Table 20-14	Examples of Effective Behaviors of Trigger Case Physician

- Introduced himself
- Asked directed questions
- Asked "pertinent negatives" (e.g., attempted to rule out pyelonephritis, checked last menstrual period)
- Obtained an abbreviated sexual history (often avoided on the phone)
- Made accurate diagnosis
- Attempted to address patient's hidden concerns
- Offered choice of plan (e.g., emergency department or wait until morning)
- Confirmed that he understood patient's choice of action (e.g., go to emergency department)

Discussion of Ineffective Behaviors

When discussing ineffective behaviors, the group should be as specific as possible and describe the physician's behaviors, not the participants' interpretations. For example, saying "The physician was uncaring" is a conclusion that cannot be verified. However, describing the specific behaviors that prompted the participant to draw this conclusion will be more helpful in understanding how to practice telephone medicine better. In this case, the physician may have appeared uncaring when he ineffectively addressed the patient's obviously worried state. His attempt to reassure her that she didn't have cancer when it was unclear if this was even a concern of hers is an example of inappropriate reassurance. This inappropriate reassurance may have resulted from an incomplete assessment of the patient's concerns. By exploring the behavior that gave the impression that the physician was uncaring, the participants can fully appreciate what skills must be developed to effectively care for patients over the telephone, such as exploring the patient's specific worries and concerns and effective patient education and reassurance. Table 20-15 lists some other ineffective behaviors in the trigger case included in Appendix I. The "alternative ending" can be used as an example of how the call changed in a positive way after the resident called his attending.

As the list of ineffective behaviors grows, the group's discussion of what are important behaviors in a telephone interview should become

---

Table 20-15   Examples of Ineffective Behaviors of Trigger Case Physician

- Did not effectively handle patient's concern about "something serious" (may in fact have raised new fears in patient)
- Did not convey he understood her degree of discomfort and concern
- Advised patient to go to the emergency department rather than called in prescription (may have led to prolonged symptoms, increased cost)

---

more sophisticated. This is a good lead-in to the skills practice and role play. Additionally, this case may raise some questions about what is the best way to manage dysuria over the phone. This topic is addressed elsewhere in this book in Chapter 8.

Part 3: Skills Practice and Role Play

Objectives

Knowledge—Participants will:

- Identify gaps in their own knowledge base important for telephone management

Attitudinal—Participants will:

- Demonstrate a commitment to apply effectively the skills of the telephone interview (data gathering, rapport building, negotiation/providing information) and decision-making, as listed below and in Tables 20-3 through 20-6
- Demonstrate an understanding of the caller's perspective by role playing as the patient

Skills—Participants will:

- Practice gathering appropriate information about chief concerns, including pertinent background information (past medical history, medications, social history, etc.)
- Practice developing rapport over the telephone

- Practice negotiation and providing education about the management plan, including home management and patient self-assessment
- Practice more advanced and specific skills as selected by individual participant
- Receive and provide useful feedback on performance

The focus of this exercise is to practice interviewing a patient on the phone and to receive feedback on the interview. The participants should generate more specific tasks for the practice depending on what they find to be the most challenging. The selected tasks can be broad, such as conducting a full interview, or narrow, such as handling a patient's anger, depending on the needs of the participants.

The two methods for performing this practice interview are role play, in which the role of the patient is played by another participant, and standardized patient interview, in which a trained actor is used to play the case. The primary advantages of each technique are summarized in Table 20-16. Depending on the goals of the participants and the organizers of the training sessions, either or both of these may be selected.

In addition to training, use of a standardized patient allows for evaluation of the participants. An observed structured clinical examination, in which the actor is trained to play the part the same every time, allows the participant's abilities to be compared to a minimum standard. The actor, or a trained observer, can look for specific behaviors given on a checklist (see Appendix III, page 410) (8).

Table 20-16   Advantages of Role Play with Physician Versus Using a Standardized Patient

Role Play (Physician Plays Part of Patient)	Standardized Patient
Participants learn patient's perspective "first hand"	Trained actor may be more realistic
Participants can select the case in the moment	Participants' performances can be measured more objectively
No additional expense in hiring/training actors	Actor may be more honest in feedback
No additional time needed for hiring/ training actors	Participants may take feedback more seriously

---

Table 20-17   Summary of Role-Play Exercise

1. Select a clinical situation and clarify the goals of the practice

2. Clarify who will be a "patient" and who will be a "doctor" in the exercise

3. Decide ground rules, including time allotted, "time outs," and sitting positions

4. Debrief beginning with the doctor, then patient, then observer/checklist

5. Repeat challenging aspects

Conducting a Role Play of a Telephone Consultation

The process for conducting this exercise is summarized in Table 20-17 and discussed in detail below.

Select a Clinical Situation and Clarify the Goals of the Practice

The success of the role play in promoting learning of telephone medicine hinges on the proper selection and elaboration of the case. This stems from two principles: physicians, especially residents, are often resistant to role play, and the scope of clinical content is so wide that pre-prepared cases cannot possibly predict the group's primary needs. The selection of the case must therefore take full consideration of the participants' most acute challenges. The exception to this is learners at the introductory stage: medical students or interns generally benefit from more generic cases in which all aspects of the interview are touched on at a basic level.

Appropriate cases for more experienced physicians can be generated in two ways:

1. *Spontaneous case selection:* Participants can be asked in the moment, "What type of case do you find most challenging?" The facilitator can also raise some ideas generated from the earlier discussion of challenges in telephone medicine.

2. *Pre-determined by facilitator/surveys:* A facilitator who is well aware of the telephone challenges experienced by the participants may write up some cases ahead of time. This is most likely to represent the group's needs if a survey has been distributed before the training session. (Four sample scenarios are given in

Appendix II, page 404.) However the cases are chosen, the group should agree on what the essential goal or task to be completed in the interview is. This may prevent unproductive sidetracking.

Clarify Roles: Who Will Be a "Patient" and Who Will Be a "Doctor" in the Exercise

The participants should be dividing into pairs of "doctors" and "patients." The facilitator may ask for volunteers to be doctors and patients, but it is often more efficient to simply assign the roles based on where the participants are sitting (for example, every other person can be a doctor).

An alternate strategy is to divide the participants into groups of three or four, with the third and fourth participants serving as observers. This is especially useful when there is a larger number of participants, which prevents the facilitator from monitoring the activity of each of the groups.

The doctors and patients should be informed of the details of the respective roles, either with a written or a verbal description. Optimally, the doctors can leave the room for a few minutes while the patients discuss the finer points of their roles. The facilitator should clarify the beginning point (especially if it's not the beginning of the call) and the ending point. An option that allows for a greater degree of feedback is to include a third participant in each group to be an observer. The use of a checklist (see Appendix III) ensures that feedback will be provided on at least a minimum set of behaviors (9).

The Role Play: Ground Rules, Including Time Allotted and "Time Outs," and Sitting Positions

Ground rules are important for providing a sense of safety. This is especially important if participants have chosen to focus on particularly challenging tasks, in which they may get "stuck" or start to flounder. To prevent this from contributing to learned helplessness, the participant playing the doctor should have permission to stop the action and take a "time out" at any point. If this occurs, the participant may step out of character and ask for help from the facilitator, other participants, or the "patient." The participant can then pick up the action again, or the debriefing may begin.

The sitting positions of the doctor and patient are important in maintaining the feel of a real phone call. A low-tech approach is simply to have the doctor and patient face away from each other. Other ways to avoid visual contact is to set up a screen dividing doctor and patient or to have the patient and doctor players sit on either side of an open door. An alternative is to conduct the role play over actual phone lines. Either way, avoiding actual visual contact is essential for adequate simulation of a phone call.

Debrief Beginning with the Doctor, Then Patient, Then Observer/Checklist

The Doctor's Perspective

Adequate time for debriefing is essential to ensure that the learner absorbs and processes the experience and the feedback. This can be done in the small groups initially and then summarized in the larger group. First, the doctor should be asked about his or her own perception of the success of the interview. Start by asking the doctors, "What went well?" This encouragement of self-reflection models a good behavior that should be done after every phone consultation. Reviewing one's own performance allows a physician to reinforce good habits and identify bad ones to be avoided in the future. Asking, "What went well?" also avoids the unpleasant trap of dwelling only on the learner's weaknesses (9).

Next ask, "What did you find difficult?" Responses to this question fall into two categories: 1) tasks that were difficult but were performed competently, and 2) tasks that were not performed competently. It is worth asking the participant if he or she thinks the task was performed competently, but this question should also be put to the patient and, if available, the observer.

The Patient's Perspective

Feedback from the patient, both specific and general, can be quite valuable. Specific feedback concerning the success of meeting the primary goals/tasks designated before the role play is essential. Whenever possible, the patient should cite specific physician behavior that was instrumental, such as "When you said 'you sound worried' and asked me what was worrying me, I could open up to you." The patient should also address any concerns the doctor mentioned in debriefing and address how the physician managed any barriers raised by the limitations of the telephone. The patient may give the doctor suggestions about what alternate approach might have been more effective in dealing with a specific challenge.

General feedback should be provided as well. This should address the patient's level of satisfaction at the end of the interview and the degree to which the patient was made to feel at ease about placing the call. A helpful question to this end is, "Would you feel comfortable calling this doctor back if there were a change in your situation?"

The Observer's Perspective

The observer can begin by commenting on any concerns the physician had about his or her own performance but should try to be as objective as possible. This entails citing specific behaviors and, whenever possible, providing direct quotes. A checklist can be very useful here because it promotes a greater degree of objectivity in both positive and negative feedback. For example, "You asked the patient about allergies and past medical history and ruled out ectopic pregnancy" is more useful than "You took a good history."

Negative feedback is especially important. Because this may be the first chance the physician has had to be observed in a telephone interview, correctable behaviors must be pointed out explicitly. Because it is easy for an observer to neglect to give negative feedback, a checklist allows the observer to give the negative feedback as an observed behavior and not as a judgement of inadequacy. For example, feedback such as "You didn't tell the patient what to watch out for and when to call back" is clearly essential to any practitioner of telephone medicine. If this is presented as an item of a checklist for which the doctor completed most other tasks successfully, the feedback may be received as a specific correctable behavior and not as inadequacy. Furthermore, the physician can be given an opportunity to correct this behavior right then by restarting the role play.

Repeat Challenging Aspects

Finally, participants should be given the chance to correct any mistakes or to re-play any difficult aspect of the interview. This may be set up with the doctor again playing the doctor role, or the participants may exchange roles. The participant playing the doctor may want to decide with the group ahead of time what approach he or she will be taking. After repeating the role play, the groups should quickly debrief again, focusing on what was different and how it worked.

Part 4: Discussion of Documentation

Objectives

>*Knowledge*—Participants will:

>• List the reasons for documentation of telephone encounters, including legal aspects and continuity of care
>• Identify the necessary elements of documenting a telephone consultation

>*Attitudinal*—Participants will:

>• Express an understanding of the value of documentation of telephone encounters

Wrapping up with a brief discussion of documentation serves two purposes. First, it highlights the necessity of documentation at a point in the session when the participants are most aware of the complexity and importance of care provided over the telephone. Second, it allows for a discussion of documentation that is driven by clinical considerations, not merely legal considerations (although legal considerations should be discussed as well; see Chapter 3).

The essential questions to be answered here are listed in Table 20-18. At this point in the training session, the participants should be highly activated as a result of the skills practice and should be able to answer most of

---

Table 20-18   Topics for Documentation Discussion

1. Why is documentation of telephone encounters important?

2. What should be documented?

3. What are the barriers to documentation?

4. How can these barriers be minimized/surmounted?

---

these questions themselves. The facilitator should simply ask the questions and make a list of the responses. This also serves to develop the participants' positive attitude about documentation: they are more likely to be committed to documentation if they themselves come up with the reasons for doing it. In the author's experience, groups with even minimal experience can come up with most of the answers to the questions in Table 20-18. Tables 20-19 through 20-21 provide a framework for these discussions, but the facilitator should keep in mind that the goal here is not knowledge but a commitment to document telephone encounters. Physicians generally document phone calls at a low rate, a rate that would be considered

---

Table 20-19   Reasons to Document Telephone Encounters

- Medicolegal issues
- Communicate with primary care physician
- Demonstrate possible abuses of the system, such as repeated requests for controlled substances

---

Table 20-20   Important Items to Document in a Telephone Encounter

History
- Chief concerns and underlying/hidden concerns
- Past medical history (including medications, allergies, and surgery)
- Relevant social/sexual history including drug use, possibility of pregnancy (last period, contraception)
- Pertinent negatives for ruling out possible emergencies
- Other relevant supportive data such as family member's impression and physician's observation of how patient *sounds*

Impression
- Working diagnosis
- Whether patient needs to be seen in person, and if so, when
- Justification for ruling out more urgent problems

Management Plan
- In-person evaluation: if, where, and when
- If 911 will be called and by whom
- Home treatment: medications, fluids, etc.
- When patient should call back
- Patient's understanding of above (and how physician knows this, i.e., did patient repeat instructions/plan?)

Table 20-21   Barriers to Documentation and Possible Solutions

Barrier	*Solution*
• Too busy/too many phone calls	• Learn to document efficiently
• Not in hospital for many phone calls	• Establish system with medical records department to insert notes into chart
• No medical record available when calls come in	• Use computerized medical records system

unacceptable were it to apply to in-person encounters. If the participants increase their rate of documenting care provided over the phone as a result of this training session, that alone would justify the time spent.

FURTHER TELEPHONE TRAINING ACTIVITIES

After this skills training session, further activities may be conducted to reinforce learning and to build on it. Table 20-22 lists some possible activities. Regardless of context for these follow-up activities, the same structure of reviewing experiences, triggering case discussions, and practicing the interview-documentation discussion will likely serve the needs well. For example, in a resident morning-report setting, the residents can begin by discussing recent challenging calls (review of experiences). One case can be selected for the resident to present his or her recollection with as much detail as possible (preparation of actual tape of the dialogue is not usually feasible). The resident's approach to the challenge can be discussed (trigger case discussion). A role- play scenario can be created in the moment with a focus on effectively managing the salient challenge (practice interview). The conference can be wrapped up with a review of the utility and key components of documenting this encounter.

SUMMARY

Regardless of their level of experience, most physicians confront challenges when providing medical care over the phone. These challenges vary

---

Table 20-22   Possible Follow-Up Sessions for Telephone Medicine Training

Case Review
- Morning report: review previous day's calls
- Telephone medicine case conference

Skills Assessment/Feedback
- Audiotape review of actual calls
- Feedback from patients after calls
- Standardized patient/observed structured clinical exam

---

greatly depending on the setting in which care is provided, the level of physician experience, and the physician's strengths and weaknesses. A structured telephone medicine curriculum can assess the physician's perceived needs and provide learning opportunities to develop the knowledge, skills, and attitudes relevant to those needs.

APPENDIX I

(Written by Anna B. Reisman)

TRIGGER TAPE SCRIPT: DYSURIA

Characters
Operator
Maria Tyler, the patient
Dr. Amis, the physician

Operator's voice: Answering service, may I help you?
Maria Tyler: I'd like to speak to Dr. Berman, please.
Operator: He's not on call today. I can take your name and number and have the covering physician call you back.
Maria: Okay, my name is Maria Tyler. My phone number is 755-1871.
Operator: What is the problem, Ms. Tyler?
Maria: I'm having pain whenever I go to the bathroom.
Operator: I'll let the doctor know right away. He should return your call shortly.

The operator pages Dr. Amis, a 30-year-old medical resident. He checks the number on his beeper and calls the operator.

Dr. Amis: Hello. This is Dr. Amis returning a page.

Operator: Dr. Amis? One of Dr. Berman's patients has just called for you, Maria Tyler. Her number is 755-1871. She has pain when she goes to the bathroom.

Dr. Amis: Thanks. (Hangs up and dials the number).

Maria's phone rings and she picks it up.

Maria: Hello?

Dr. Amis: Maria Tyler, please.

Maria: Speaking.

Dr. Amis: This is Dr. Amis. I'm covering for Dr. Berman. What can I do for you?

Maria: Oh, Doctor, I'm having terrible pain when I go to the bathroom ever since this morning, and I saw some blood in it!

Dr. Amis: Are you talking about when you urinate or when you have a bowel movement?

Maria: When I urinate.

Dr. Amis: What kind of pain is it? Pressure, burning?

Maria: Oh, it burns! And I have to keep going every few minutes, and only a little comes out.

Dr. Amis: Did you ever have this problem before?

Maria: Uhh, about two years ago I had a bladder infection that was kind of like this pain, but I never saw any blood!

Dr. Amis: Do you have any back pain? Fevers? Any discharge?

Maria: No.

Dr. Amis: Nausea? Vomiting?

Maria: No, no.

Dr. Amis: Any other medical problems? Do you take any medications?

Maria: No.

Dr. Amis: How old are you?

Maria: Twenty-eight.

Dr. Amis: When's the last time you had sex?

Maria: Oh, about a couple days ago, I guess.

Dr. Amis: Last period?

Maria: Uh, about two weeks ago. Is there something you can give me, doctor? It really hurts!

Dr. Amis: Well, I'm pretty sure that you have another bladder infection. You'll need to have a urine test done to make sure, so you can either go to an emergency department tonight or wait until the morning and go in to see Dr. Berman. How does that sound?

Maria: Uh, okay, I guess. You really think it's just a bladder infection?

Dr. Amis: Chances are pretty good. Is there something else you're worried about?

Maria: Well, I'm concerned about the blood. Are you sure it couldn't be something else?

Dr. Amis: Well, it's pretty common for bladder infections to cause some blood in the urine, so it's probably from that. I don't think you have cancer, if that's what you're worried about. You're really too young for that.

Maria: Oh.

Dr. Amis: So, that's it, unless you have any more questions. Do you think you'll go to the emergency department now, or wait until tomorrow?

Maria: I think I'll go to the emergency department now. I won't be able to sleep with this pain otherwise!

Dr. Amis: Okay. Good luck, I hope you feel better.

Maria: Thanks, doctor.

Alternative Ending

Dr. Amis: So, that's it. If you don't mind, I'm just going to check something and I'll call you right back.

Maria: Okay.

Dr. Amis calls his back-up attending for a 3-minute discussion of the case and the plan, then calls Maria back. In the discussion, the attending realizes that because the patient has no comorbid illnesses and a history of previous uncomplicated UTI, she can be offered empiric treatment.

Dr. Amis: Hi, Ms. Tyler?

Maria: Yes.

Dr. Amis: Actually, I think I can spare you a trip to the emergency department and just give you some antibiotics. I can call it in to your pharmacy if you give me the number.

Maria: Great! Just a moment...okay, it's 544-8938.

Dr. Amis: And, I forgot to ask, do you have any allergies?

Maria: Um, I think to sulfa.

Dr. Amis: Okay. I can also give you a medication to help ease the burning. It may make your urine orange. Please be sure to call the office tomorrow and make a follow-up appointment with Dr. Berman.

Maria: Okay, doctor. Thanks so much.

APPENDIX II

(Written by Anna B. Reisman)

CASE SCENARIOS FOR ROLE PLAY

Case 1   Sore Throat

Time: Saturday at 3 p.m.
Patient: Paul Levi
Age: 22

Operator: Doctor, you have a call from one of Dr. Porter's patients, Paul Levi, who's having a sore throat. The phone number is 555-1515.

Background Information for Learner Playing the Part of Patient

You have a very big date tonight, and you woke up this morning feeling sick. Your main concerns revolve around the date: feeling well, not looking sick, and not being contagious. You're a little embarrassed about calling because you know you're not that sick, but you want to get better as quickly as possible. You think the best thing would be for you to get antibiotics. Your roommate had the same thing, and antibiotics seemed to work.

Chief complaint: "I'm having a pretty bad sore throat. I think I need antibiotics."

Answer only if asked:
- Throat feels "scratchy," worse with swallowing, but not bad
- Symptoms started this morning
- No fever/sweats/chills
- Nose is running, clear discharge
- Roommate has "the same thing (he/she got antibiotics two days ago and is now better!)"
- No chronic medical problems/medications/allergies
- (If female) Using oral contraceptive pills

Underlying concerns
- "Big date" tonight: it's important that you feel and look well for it
- "Can I give this to someone else by kissing?"

Physician Checklist

	Yes	No
A. History Taking		
1. When symptoms started	___	___
2. Pain with swallowing?	___	___
3. Sneezing or rhinorrhea?	___	___
4. Painful neck?	___	___
5. Myalgias?	___	___
6. Fever?	___	___
7. Other people with similar symptoms?	___	___
8. Why such a desire for antibiotics?	___	___
9. Allergies?	___	___
10. Meds?	___	___
B. Management		
1. Prescribes analgesic for pain	___	___
2. Recommends throat culture the following day	___	___
3. Recommends fluids and rest	___	___
4. Discusses lack of need for antibiotics at this time	___	___

Case 2   Diarrhea

Time: Tuesday, 2 a.m.
Patient: Henry Firestone
Age: 39

Operator: Hello, Doctor. One of Dr. Isaac's patients wants to talk to you, Henry Firestone. He has diarrhea. The phone number is 555-1616.

Background Information for Learner Playing the Part of Patient

You've had diarrhea since this morning, and it's pretty bothersome; you've had to leave work. You think it's probably nothing to worry about, but in the back of your head you're worried it could be your appendix. Your son had his appendix taken out 2 weeks ago, and it nearly ruptured. You're also nervous about missing work and about whether you can go out for your anniversary tomorrow night.

Chief complaint: "Doctor, I've got the runs. It won't stop."

Answer only if asked:
- Started this morning while at work, left work at noon
- Diarrhea is brown, watery, no blood, 4 to 5 bowel movements today
- Mild crampy pain
- Don't feel feverish, but no thermometer
- Nauseated, but no vomiting
- Have very little appetite, but can drink water
- No other symptoms
- No chronic medical problems
- On no medications, do not use alcohol or drugs
- No allergies
- No recent travel
- (If female) Have had "tubes tied"

State if asked for other questions or concerns:
- You're very worried about appendicitis. Your son just had it a few weeks ago, and you still have your appendix.
- Should you go to work? (You work as a chef.)

- You have theater tickets for tomorrow, and you don't want to keep running to the bathroom! Should you take any medicines?

Physician Checklist

		Yes	No
A.	History Taking		
	1. When symptoms started	___	___
	2. Quality of diarrhea/number of stools	___	___
	3. Vomiting?	___	___
	4. Fever or dizziness?	___	___
	5. Abdominal pain?	___	___
	6. Past medical history?	___	___
	7. Allergies/meds/pregnant?	___	___
	8. Any other concerns?	___	___
	9. Why so worried about appendicitis?	___	___
B.	Management		
	1. Explains that probable self-limited illness	___	___
	2. Explains symptoms of appendicitis	___	___
	3. Recommends fluids and rest	___	___
	4. Recommends symptoms relief (e.g., Pepto-Bismol)	___	___
	5. Recommends call back if fever or severe pain or bleeding develops	___	___
	6. Recommends follow-up appointment in a few days if not improved	___	___

Case 3 Abnormal Laboratory Result: Cholesterol

Time: Monday, 4:30 p.m.
Patient: Mary Creighton
Age: 42

Background Information for Learner Playing the Part of Physician

You make a call to Mary Creighton, a 42-year-old woman who had bloodwork done several days ago at her last visit with you. She is a smoker and has hypertension. Her labs were all normal except for her lipid profile, with a total cholesterol of 260, an HDL of 35, and an LDL of 170.

Background Information for Learner Playing the Part of Patient

You smoke one half of a pack of cigarettes a day, and you have no interest in quitting. You have hypertension and take atenolol and hydrochlorothiazide for it. You saw the doctor last week and had fasting bloodwork.

Only mention if asked:
You've seen the ads for some cholesterol-lowering medications and have heard that they can have serious side effects. You are very reluctant to take such medication.

Your diet is fast-food cheeseburgers two or three times a week for lunch, eggs for breakfast almost every day, red meat for dinner several times a week.

Physician's Checklist

	Yes	No
1. Gave lab results and explained implications	___	___
2. Checked patient's understanding	___	___
3. Checked and addressed patient's concerns	___	___
4. Asked about diet	___	___
5. Gave dietary information on phone and/or suggested in-person visit for this	___	___
6. Asked if patient had any questions	___	___

Case 4   Headache: Patient Expects a Prescription to be Called In

Time: Sunday (on a 3-day weekend) at noon
Patient: Nancy Gifford
Age: 35

Operator: Doctor, you received a call from one of Dr. Porter's patients, Nancy Gifford, who has a severe headache. The phone number is 555-1414.

Background Information for Learner Playing the Part of Physician

You are not the regular physician of this patient.

Background Information for Learner Playing the Part of Patient

You are calling because of severe headaches and you want Tylenol with codeine, which usually helps. The other doctors on call have always called the prescription in for you when you needed it, although sometimes they seem to hesitate a bit until you make it clear that you *need* the pills. You are very impatient with the doctor on the phone, who asks you a lot of questions when all you need is the prescription.

Chief complaint: "I'm having one of my headaches again, and I need you to call in a prescription for me."

Only state if asked:
- You have had bilateral temporal throbbing and nausea for a few hours.
- This is typical for your headaches.
- You have *none* of the following: recent head trauma, "worst headache in your life," fever, stiff neck, vomiting, other medical problems.
- Tylenol with codeine is the only thing that has helped.
- You've tried ibuprofen and it never helped.
- You have never tried any other medications.
- You've never been given a diagnosis for the headaches other than "bad headaches."
- You last saw your regular doctor in person about 6 months ago.
- You last had one of these headaches a couple of weeks ago.
- They usually come on every month or so.

Physician Checklist

	Yes	No
A. History Taking		
1. Asked for description of headaches	___	___
2. Asked if this is typical for headaches, or not	___	___
3. Asked about previous medications tried	___	___
4. Asked when last saw regular physician	___	___
5. Asked about fever/stiff neck/nausea/vomiting/ if "worst headache ever"	___	___
6. Asked if history of cancer or HIV	___	___
B. Management		
1. Encouraged patient to seek refills during regular office hours in the future	___	___
2. Encouraged patient to see regular physician soon to have headaches evaluated	___	___
3. Offered suggestions for non-pharmacologic or non-narcotic treatments for the headache	___	___

C. If Physician Feeling Pressured to Write Prescription   Yes   No

    1. Told patient feeling pressured, and told patient   ___   ___
       concerned about using addictive medications
       improperly

D. If Physician Decides to Prescribe Medicine

    1. Told patient he/she needed to verify information   ___   ___
       offered by patient (from pharmacy or patient's
       regular physician or medical chart)

    2. Prescribed smallest quantity possible   ___   ___

APPENDIX III

CHECKLIST FOR SKILLS PRACTICE

The following checklist contains the basic tasks in assessing a problem over the phone. The individual cases in Appendix II (pages 404-410) have additional checklists with tasks relevant to the specific cases.

Checklist for Skills Practice

	Yes	No	N/A
I. History Taking			
A. Opening			
1. Doctor's name	___	___	___
2. Nurse's/receptionist's name	___	___	___
3. Caller's name	___	___	___
4. Patient's name	___	___	___
5. Age/sex	___	___	___
6. Personal physician	___	___	___
B. Take appropriate clinical history			
1. Duration of problem	___	___	___
2. Details of symptoms	___	___	___
3. Previous history of problem	___	___	___
4. Significant past medical history	___	___	___
5. Anything done to relieve symptoms	___	___	___
C. Determine actual reason for calling if different			
from presenting problem			
1. Uncover fears about problem	___	___	___
2. Assess emotional state of caller	___	___	___
3. Establish urgency of caller's needs	___	___	___

Checklist for Skills Practice (cont'd)

	Yes	No	N/A
I. History Taking (cont'd)			
D. Obtain psychosocial history if appropriate			
1. Current occupation	—	—	—
2. Living situation	—	—	—
3. Recent changes/stress	—	—	—
4. Support systems	—	—	—
II. Decision Making			
A. Use appropriate screening questions	—	—	—
B. Stay open to new information	—	—	—
C. State opinion about nature of problem	—	—	—
III. Management			
A. Make appropriate medical assessment			
1. Recapitulate chief complaint	—	—	—
2. State opinion about seriousness	—	—	—
B. Suggest a plan for management			
1 Establish feasibility of plan	—	—	—
C. Offer appropriate advice and reassurance	—	—	—
IV. Contracting/Compliance			
A. Negotiate with caller on management			
1. Agree on plan with caller	—	—	—
B. Make definite statement about follow-up	—	—	—
C. Offer appropriate patient education			
1. About the problem	—	—	—
2. About the treatment	—	—	—
D. Offer opportunity for further questions	—	—	—
V. Communication Skills			
A. Use speaking skills	—	—	—
1. Voice is clear	—	—	—
2. Tone of voice implies interest	—	—	—
B. Use appropriate questioning style			
1. Organize questions	—	—	—
2. Avoid over-explanation/jargon	—	—	—
3. Help caller focus on problem	—	—	—
4. Use open-ended questions	—	—	—
5. Avoid premature closure	—	—	—
6. Allow time for caller to tell story	—	—	—
7. Use telephone time efficiently	—	—	—
8. Avoid causing anxiety	—	—	—
9. Use appropriate reassurance	—	—	—

Modified from Curtis P, Evens S. Telephone Medicine: Instructor Manual. Chapel Hill, NC: Health Sciences Consortium; 1983.

REFERENCES

1. **Hannis MD, Hazard RL, Rothschild M, et al.** Physician attitudes regarding telephone medicine. J Gen Intern Med. 1996;11:678-83.
2. **Curtis P, Talbot A.** The after-hours call in family practice. J Fam Pract. 1979;9: 901-9.
3. **Flannery MT, Moses GA, Cykert S, et al.** Telephone management training in internal medicine residencies: a national survey of program directors. Acad Med. 1995;70:1138-41.
4. **Fleming MF, Skochelak SE, Curtis P, Evens S.** Evaluating the effectiveness of a telephone medicine curriculum. Med Care. 1988;26:211-6.
5. **Ottolini MC, Greenberg L.** Development and evaluation of a CD-ROM computer program to teach residents telephone management. Pediatrics. 1998;101:461.
6. **Smith SR, Fischer PM.** Patient management by telephone: a training exercise for medical students. J Fam Pract. 1980;10:463-6.
7. **Skeff KM, Stratos GA, Berman J.** Educational theory and teaching medical interviewing. In: Lipkin M, Putnam SM, Lazare A, eds. The Medical Interview. New York: Springer-Verlag; 1995:379-87.
8. **Curtis P, Evens S.** Telephone Medicine: Instructor and Participant Manual. Chapel Hill, NC: Health Sciences Consortium; 1983.
9. **Cohen-Cole S, Bird J, Mance R.** Teaching with role play: a structured approach. In: Lipkin M, Putnam SM, Lazare A, eds. The Medical Interview. New York: Springer-Verlag; 1995:405-12.

INDEX